Monographs in Computer Science

Editors

David Gries
Fred B. Schneider

Springer-Verlag Berlin Heidelberg GmbH

Monographs in Computer Science

Abadi and Cardelli, **A Theory of Objects**

Benosman and Kang [editors], **Panoramic Vision: Sensors, Theory, and Applications**

Broy and Stølen, **Specification and Development of Interactive Systems: FOCUS on Streams, Interfaces, and Refinement**

Brzozowski and Seger, **Asynchronous Circuits**

Cantone, Omodeo, and Policriti, **Set Theory for Computing: From Decision Procedures to Declarative Programming with Sets**

Castillo, Gutiérrez, and Hadi, **Expert Systems and Probabilistic Network Models**

Downey and Fellows, **Parameterized Complexity**

Feijen and van Gasteren, **On a Method of Multiprogramming**

Herbert and Spärck Jones [editors], **Computer Systems: Theory, Technology, and Applications**

Leiss, **Language Equations**

McIver and Morgan [editors], **Programming Methodology**

Misra, **A Discipline of Multiprogramming: Programming Theory for Distributed Applications**

Nielson [editor], **ML with Concurrency**

Paton [editor], **Active Rules in Database Systems**

Selig, **Geometrical Methods in Robotics**

Shasha and Zhu, **High Performance Discovery in Time Series: Techniques and Case Studies**

Dennis Shasha
Yunyue Zhu

High Performance Discovery In Time Series

Techniques and Case Studies

Springer

Dennis Shasha
Courant Institute
251 Mercer Street
New York, NY
USA
shasha@cs.nyu.edu

Yunyue Zhu
Courant Institute
251 Mercer Street
New York, NY
USA
yunyue@cs.nyu.edu

Series Editors:
David Gries
Department of Computer Science
Cornell University
4130 Upson Hall
Ithaca, NY 14853-7501
USA

Fred B. Schneider
Department of Computer Science
Cornell University
4130 Upson Hall
Ithaca, NY 14853-7501
USA

ISBN 978-1-4419-1842-0 ISBN 978-1-4757-4046-2 (eBook)
DOI 10.1007/978-1-4757-4046-2

Cover image designed by Cloe L. Shasha

Printed on acid-free paper.

To Victor and Angela, for their inspirational music. – *D.S.*

To my parents and Amy, for many wonderful things in life. – *Y.Z.*

Preface

Overview and Goals

Data arriving in time order (a data stream) arises in fields ranging from physics to finance to medicine to music, just to name a few. Often the data comes from sensors (in physics and medicine for example) whose data rates continue to improve dramatically as sensor technology improves. Further, the number of sensors is increasing, so correlating data between sensors becomes ever more critical in order to distill knowledge from the data. On-line response is desirable in many applications (e.g., to aim a telescope at a burst of activity in a galaxy or to perform magnetic resonance-based real-time surgery). These factors – data size, bursts, correlation, and fast response – motivate this book.

Our goal is to help you design fast, scalable algorithms for the analysis of single or multiple time series.

Not only will you find useful techniques and systems built from simple primitives, but creative readers will find many other applications of these primitives and may see how to create new ones of their own. Our goal, then, is to help research mathematicians and computer scientists find new algorithms and to help working scientists and financial mathematicians design better, faster software.

Organization and Features

The book is divided into two parts. Part I describes fast algorithms for a few powerful primitives, building up the concepts from a tutorial review of underlying mathematical concepts. Part II integrates and extends those primitives into useful applications ranging from music to physics, finance and medicine.

The primitives can be broken down into three major categories:

- Fast transformation algorithms (Fourier, Wavelet, Singular Value Decomposition, and Random Projection) for reducing the size of time series data by exploiting the redundancy that is often present in that data;

- Framework for comparing time series using those reductions and single and multi-dimensional indexes (such as B-trees, R-trees, and Grid files); and,
- Methods for increasing the flexibility of similarity searching (shifting, scaling, and warping).

These techniques are the basis for fast primitives that can speed up many time series matching applications from stock market correlation to musical matching. Whenever there is a choice of techniques, we suggest decision criteria.

Part II describes three case studies based on and extending the primitives of Part I. (We ask you to take particular note of the extensions as they come up. Getting the most use out of this book will require your creative use of the ideas presented here.)

- *StatStream* Given a window size w (say an hour), a small offset b (say a minute) and a streaming set of N (in the thousands or more) time series, StatStream will find at time $w + t + b$ the most highly correlated d pairs among the N based on the window between t and $t + w$. A short b time later, the system will find the most highly correlated pairs in the window between $t + b$ and $t + w + b$. Thus, StatStream finds the highly correlated pairs in a "punctuated" continuous way. We illustrate an application on stock prices, but you may find many other natural applications. In fact, the speed of this approach can make some algorithms real-time that are now off-line.
- *HumFinder* Most people (including the authors) hum rather poorly. We don't sing at the correct absolute pitch. We don't even sing the intervals between the pitches correctly. Our tempo is imprecise, sometimes too slow and sometimes too fast. Nevertheless, somehow people have some idea as to what we are humming when we hum. The goal of a query by humming program is to make a computer do what a person can do. Our software HumFinder, which we describe, uses pitch translation, time dilation, and dynamic time warping to match an inaccurate hum to a melody remarkably accurately.
- *OmniBurst* Burst detection is the activity of finding abnormal aggregates in data streams. Such aggregates are based on sliding windows over such streams. In some applications, we may not know the proper sliding window size to examine, so we may want to look at many (thousands or more) at a time. OmniBurst, software for burst detection, can detect bursts of varying duration (from a few milliseconds to several hours), stock market price volatility, and bursts in space. The software makes use of a shifted binary tree structure to create a linear time filter that can guarantee no bursts will be missed at the same time that it guarantees (under a reasonable statistical model) the filter eliminates nearly all false positives.

Audience

The book is written at a first year graduate level in computer science, physics, or mathematics and could be used as part of a course on time series or as a self-study text. We try to keep the mathematical prerequisites reasonable, assuming only that you are familiar with first year calculus and some linear algebra. Most of the more mathematical sections can be skipped on a first reading. We direct the reader to subsections labeled "The Bottom Line."

High performance time series data analysis is essential to many fields such as finance, medicine, image analysis, speech recognition, telecommunications (internet traffic), astrophysics, and potentially genomics and proteomics. Professionals and researchers who want to analyze and make sense of massive time series data may find the descriptions of the tools most pertinent.

Acknowledgments

Rich interactions with colleagues improve research and make it enjoyable. Among those who work on kindred problems, we would like to thank Eamonn Keogh for the UCR time series data mining archive and Dominic Mazzoni for signal processing software. Allen Mincer has both introduced us to high-energy physics and arranged our access to Milagro data and software. Stuart Lewis has helped with many exciting ideas and promising introductions to the Magnetic Resonance Imagery community. Within our group, Richard Cole, Tony Corso, Hsiao-Lan Hsu, Flip Korn, Alberto Lerner, Nicolas Levi, S. Muthukrishnan, David Rothman, David Tanzer, Aris Tsirigos, Zhihua Wang, Xiaojian Zhao have lent both voices and helpful suggestions in the course of this work. Karen and Cloe lent only their voices, but theirs are really good. Amy Chen has proofread much of the manuscript. This research was implemented in the K language without whose elegance it would have gone much more slowly; so, heartfelt thanks to Arthur Whitney. We are very grateful for all the help provided by Springer, and thank in particular Ann Kostant, Wayne Wheeler, Elizabeth Loew and Martin Stock.

Dennis Shasha
Yunyue Zhu

New York, August 2003

Contents

High Performance Discovery in Time Series

Part I

Review of Techniques

1

Time Series Preliminaries

A time series is a sequence of recorded values. These values are usually real numbers recorded at regular intervals, such as yearly, monthly, weekly, daily, and hourly. Data recorded irregularly are often interpolated to form values at regular intervals before the time series is analyzed. We often represent a time series as just a vector of real numbers.

Time series data appear naturally in almost all fields of natural and social science as well as in numerous other disciplines. People are probably most familiar with financial time series data. Figure 1.1 plots the daily stock prices and transaction volumes of IBM in the first month of 2001.

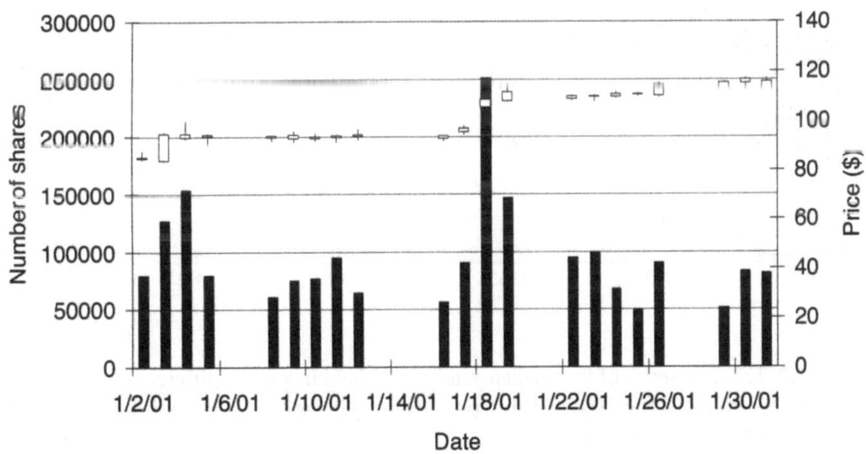

Figure 1.1. Time series of the daily open/high/low/close prices and volumes of IBM's stock in January 2001.

Figure 1.2 shows the median yearly annual household income in regions from the United States from 1975 to 2001. Economists may want to identify the trend of changes in annual household income over time. The relationship between different time series such as the annual household incomes time series from different regions are also of great interest.

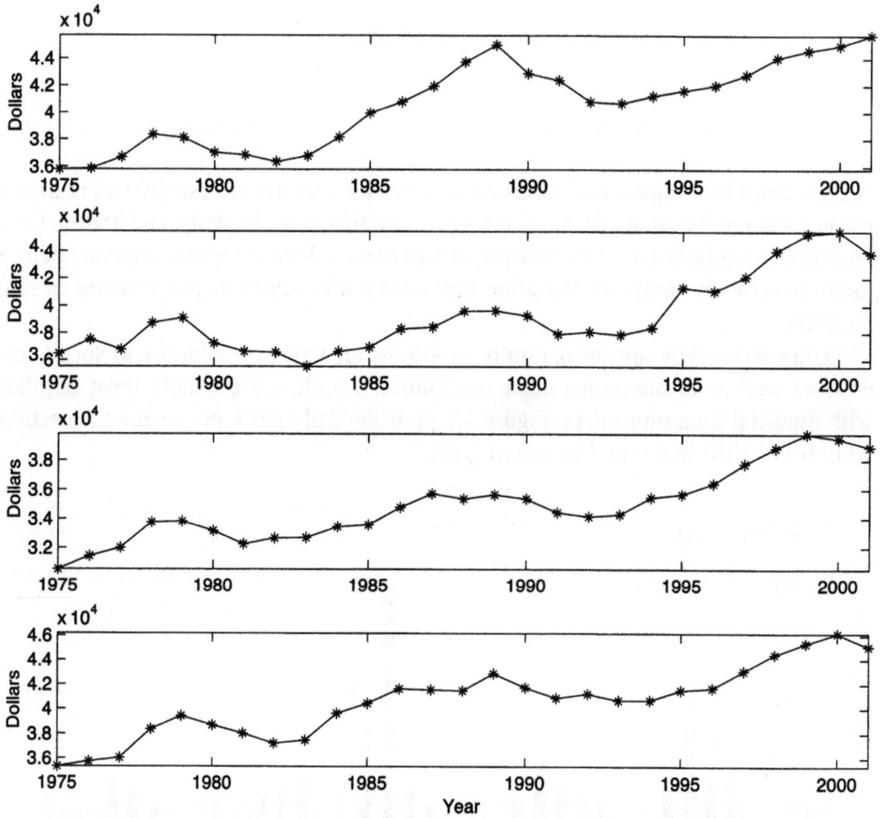

Figure 1.2. The time series of the median yearly household income in different regions of the United States from 1975 to 2001. From top to bottom: Northeast, Midwest, South and West. Data Source: US Census Bureau.

In meteorological research, time series data can be mined for predictive or climatological purposes. For example, Figure 1.3 shows the monthly average temperature in New York.

Time series data may also contain valuable business intelligence information. Figure 1.4 shows the number of bytes that flow through an internet router. The periodic nature of this time series is very clear. There are seven equally spaced spikes that correspond to seven peaks in Internet traffic over the course of a day. By analyz-

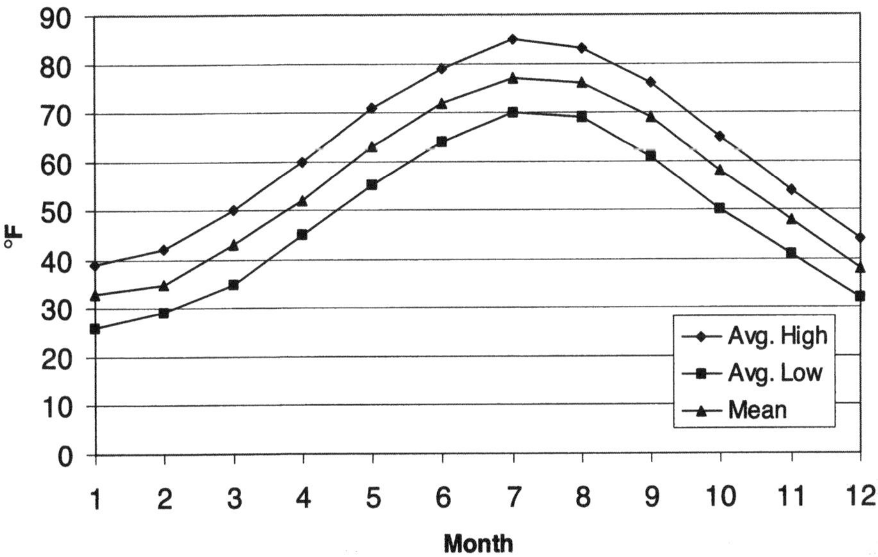

Figure 1.3. The time series of the monthly average temperature for New York. Data Source: Weather.com.

ing such traffic time series data [63, 62], an Internet service provider (ISP) may be able to optimize the operation of a large Internet backbone.

In fact, any values recorded in time sequence can be represented by time series. Figure 1.5 gives the time series of the number of Hyde Park purse snatchings in Chicago.

1.1 High Performance Time Series Analysis

People are interested in time series analysis for two reasons:

1. **Modeling time series:** to obtain insights into the mechanism that generates the time series.
2. **Forecasting time series:** to predict future values of the time series variable.

Traditionally, people have tried to build models for time series data and then fit the actual observations of sequences into these models. If a model is successful in interpreting the observed time series, the future values of time series can be predicted provided that the model's assumptions continue to hold in the future.

As a result of developments in automatic massive data collection and storage technologies, we are living in an age of data explosion. Many applications generate massive amounts of time series data. For example,

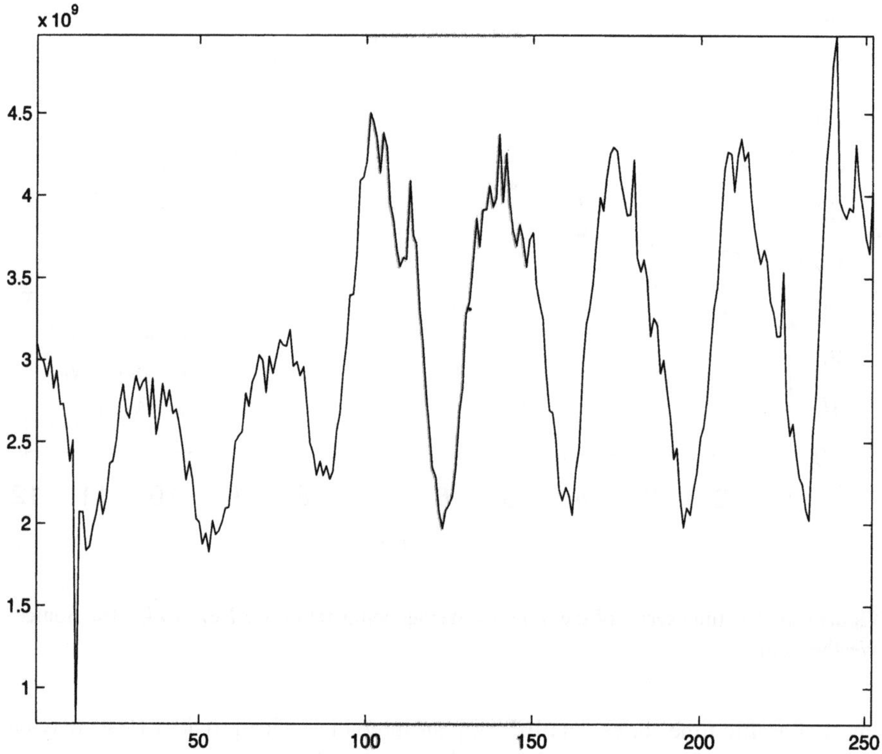

Figure 1.4. The time series of the number of bits received by a backbone internet router in a week.

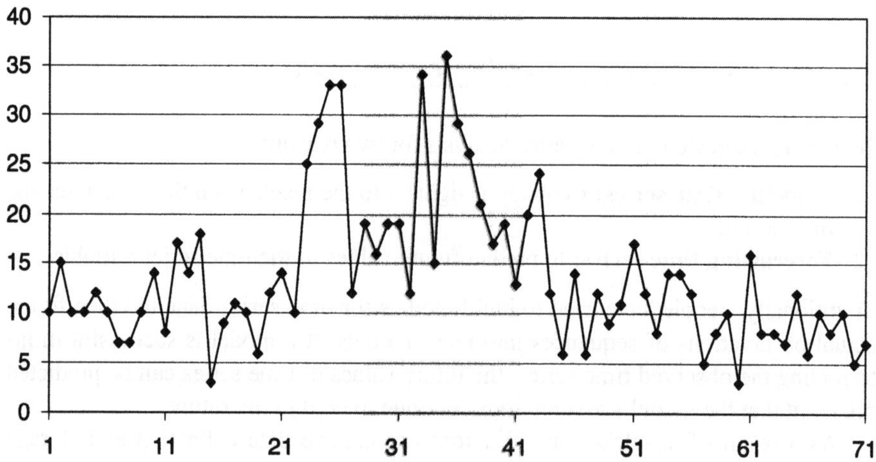

Figure 1.5. Series of the number of Hyde Park purse snatchings in Chicago within every 28-day period; January 1969 – September 1973. Data Source: McCleary & Hay (1980).

- In mission operations for NASA's Space Shuttle, approximately 20,000 sensors are telemetered once per second to Mission Control at Johnson Space Center, Houston [59].
- In telecommunications, the AT&T long distance data stream consists of approximately 300 million records per day from 100 million customers [26].
- In astronomy, the MACHO Project to investigate the dark matter in the halo of the Milky Way monitors photometrically several million stars [2]. The data rate is as high as several Gbytes per night.
- There are about 50,000 securities traded in the United States, and every second up to 100,000 quotes and trades are generated [5].

As extremely large data sets grow more prevalent in a wide variety of settings, we face the significant challenge of developing more efficient time series analysis methods. To be scalable, these methods should be linear in time and sublinear in space. Happily, this is often possible, as shown below.

1. **Data Reduction and Transformations** Because time series are observations made in sequence, past data is often a good predictor of future values. This fact gives one the opportunity to reduce the size of the data without substantial loss of information. Data reduction is often the first step to tackling massive time series data because it will provide a synopsis of the data. That synopsis in turn will lead to faster processing. Several data reduction techniques can be used for time series data and we present the ones that have been most useful in our experience. The Discrete Fourier Transform (DFT) is a classic data reduction technique. Based on the DFT, researchers have more recently developed the Discrete Wavelet Transform. Also, Singular Value Decomposition based on traditional Principal Components Analysis is an attractive data reduction technique because it can provide optimal data reduction in some circumstances. Random projection of time series has great promise and yields many nice results because it can provide approximate answers with guaranteed bounds of errors. We will discuss these techniques one by one in Chapter 2. The chapter is self-contained, but is not easy reading, so be well rested when you attack it.
2. **Indexing Methods** To build scalable algorithms we must avoid a brute force scan of the data. Indexing methods provide a way to organize data so that the data with the interested properties can be retrieved efficiently. The indexing method also organizes the data in a way so that the I/O cost can be greatly reduced. This is essential to high performance discovery from time series data. Indexing methods are the topic of Chapter 3.
3. **Flexible Similarity Searching** To discover patterns from time series, data analysts must be able to compare time series in a scale- and magnitude-independent way. Hence, shifting and scaling of the time series amplitude, up- and down-sampling for uniform changes in tempo, and dynamic time warping of time series are some useful techniques. They will be discussed in Chapter 4.

2

Data Reduction and Transformation Techniques

A line is the shortest route between two points but the easiest path wanders. – *Kentucky Pete, rock climber*

From a data mining point of view, time series data has two important characteristics:

1. **High Dimensional** If we think of each time point of a time series as a dimension, a time series is a point in a very high dimensional space. A time series of length 1000 corresponds to a point in a 1000-dimensional space. Though a time series of length 1000 is very common in practice, processing in a 1000-dimensional space is extremely difficult even with modern computer systems.
2. **Temporal Order** Fortunately, the consecutive values in a time series are related because of the temporal order of a time series. For example, for financial time series, the differences between consecutive values will be within some predictable threshold most of the time. This temporal relationship between nearby data points in a time series produces some redundancy, and such redundancy provides an opportunity for data reduction.

Data reduction [14] techniques reduce massive data into a manageable synoptic data structure while preserving the characteristics of the data to a large extent. Most data reduction techniques for time series results in reducing the dimensionality of the time series. In this book, we will focus on such data reduction methods, dimensionality reduction. There are some data reduction methods, for example, sampling methods that reduce the data size but not the data dimensionality. However, they are not as interesting in the setting of time series data mining.

The benefits of data reduction go way beyond storage reduction. The main benefit, in fact, is time reduction. For example, computing the correlation between two length n time series requires $O(n)$ time. If the lengths can be transformed to a far smaller number k, then the time goes down to $O(k)$. But the benefits don't stop there. If, as is often the case, we want to find the pairs among M streams whose correlations are close to 1, there are potentially $O(M^2)$ correlation computations to perform. Correlation is closely related to (normalized) Euclidean distance, so a multi-dimensional index could in principle help. If n is over 20 however, a multi-dimensional index does not in practice work well. (Such structures are best used for 4–12 dimensions.) On the other hand, if our data reduction technique can reduce n to k and $k < 20$, we can use an index on the reduced data, potentially reducing the work of finding high correlations to something like $O(kM)$ or perhaps $O(kM\ log\ M)$ once the data reductions are done. The entire computation would then be dominated by the time it takes to compute the data reduction. Often this time can be made linear or perhaps better. So, potentially, the entire computation time is reduced by a factor of n. This can amount to a factor of 100 or more. The easiest path wanders ....

With this motivation, we will discuss data reduction and transformation techniques for time series in detail. We start in Section 2.1 with the Fourier Transform, which has long been used in the data mining community and continues to enjoy wide use for data that tends to be periodic (e.g., sales data). Wavelet Transforms in Section 2.2 are a philosophically similar but technically different Transform technique that can be used on data with episodic (e.g., counts of alarms) as opposed to periodic structure. Fourier and Wavelet transforms are both based on orthogonal function families. Singular value decomposition (Section 2.3) is provably the optimal data reduction technique based on orthogonal function analysis. In Section 2.4, we will discuss a relatively new data reduction technique: random projection. This projection is well suited to time series that individually appear random but that may lie in non-random relation to other time series. We conclude this chapter with a detailed comparison of these data reduction techniques.

The chapter contains substantial mathematics, but it's what one might call "computation-generating" mathematics: many of the theorems explain how operations (e.g., transforms, convolutions, and correlations) can be computed. Thus, you can read the chapter algorithmically unless you want to use these primitives as black boxes. In that case, skip directly to the last section for help in deciding which black box to use.

2.1 Fourier Transform

Fourier analysis has been used in time series analysis ever since it was invented by Baron Jean Baptiste Joseph Fourier in Napoleonic France. Traditionally, there are two ways to analyze time series data: time domain analysis and frequency domain analysis. Time domain analysis examines how a time series process evolves through time, with tools such as autocorrelation functions. Frequency domain analysis, also known as spectral analysis, studies how periodic components at different frequencies describe the evolution of a time series. The Fourier transform (FT) is the main tool for spectral analysis in time series. Moreover, in time series data mining, the FT can be used as a tool for data reduction. The reason is that, in many applications, most coefficients of the FT have small values so that they can be ignored. When this happens, time series containing thousands of data points can be reduced to scores of coefficients.

Thus, the main virtues of the FT are:

- The FT translates a time series into frequency components.
- Often, only some of those component have non-negligible values.
- From those significant components, we can reconstruct most features of the original time series and compare different time series.

- Convolution, inner product and other operations are much faster in the Fourier domain.

In this section, we give a primer of the mathematics underlying the Fourier Transform and the operations that can be done with it. A deep understanding may help you develop new transformation techniques that apply to application domains having specialized properties.

The first few subsections are quite mathematical. If you want to know how this will help you, go to Subsection 2.1.6 with the title "The Bottom Line." If you want just enough math to understand what the computational primitives are doing, then go to Subsection 2.1.4 entitled "Discrete Fourier Transform." All of our work is based on sequences of discrete values rather than continuous functions. On the other hand, we believe the discussion of continuous functions will give you much insight, so we warmly recommend that you read straight through.

2.1.1 Orthogonal Function Families

In many applications, it is convenient to approximate a function by a linear combination of basis functions. For example, any continuous function on a compact set can be approximated by a set of polynomial functions. If the basis functions are trigonometric functions, that is, sinc and cosine functions, such a transform is called the Fourier Transform.

First, we give the formal definition of orthogonal function family. This is not only the basis of Fourier Transform, but also the basis of two important other data reduction techniques: the Wavelet Transform (WT) and Singular Value Decomposition (SVD). In our discussion, we assume a function to be a real function unless otherwise noted.

A function f is said to be *square integrable* on an interval $[a, b]$ if the following holds:

$$\int_a^b f^2(x)\,dx < \infty. \tag{2.1}$$

This is denoted by $f \in L^2([a, b])$.

Definition 2.1 (Orthogonal Function Family). *An infinite square integrable function family $\{\phi_i\}_{i=0}^{\infty}$ is orthogonal on interval $[a, b]$ if*

$$\int_a^b \phi_i(x)\phi_j(x)\,dx = 0, \qquad i \neq j, i, j = 0, 1, \ldots \tag{2.2}$$

and

$$\int_a^b |\phi_i(x)|^2\,dx \neq 0, \qquad i = 0, 1, \ldots. \tag{2.3}$$

The integral above is called the inner product of two functions.

Definition 2.2 (Inner Product of Functions). *The inner product of two real functions $f(x)$ and $g(x)$* [1] *on interval $[a, b]$ is*

$$\langle f(x), g(x)\rangle = \int_a^b f(x)g(x)\,dx. \tag{2.4}$$

The norm of a function is defined as follows.

Definition 2.3 (Norm of Function). *The norm of a function $f(x)$ on interval $[a, b]$ is*

$$||f(x)|| = \langle f(x), f(x)\rangle^{\frac{1}{2}} = \left(\int_a^b f(x)f(x)\,dx\right)^{\frac{1}{2}}. \tag{2.5}$$

Therefore, $\{\phi_i\}_{i=0}^{\infty}$ is an orthogonal function family if

$$\langle \phi_i(x), \phi_j(x)\rangle = 0, \qquad \text{for } i \neq j \tag{2.6}$$

$$\langle \phi_i(x), \phi_i(x)\rangle = ||\phi_i(x)||^2 \neq 0. \tag{2.7}$$

Also, if $||\phi_i(x)||^2 = 1$ for all i, the function family is called *orthonormal.*

The following theorem shows how to represent a function as a linear combination of an orthogonal function family.

Theorem 2.4. *Given a function $f(x)$ and an orthogonal function family $\{\phi_i\}_{i=0}^{\infty}$ on interval $[a, b]$, if $f(x)$ can be represented as*

$$f(x) = \sum_{i=0}^{\infty} c_i\phi_i(x), \tag{2.8}$$

where $c_i, i = 0, 1, \ldots,$ are constants, then $c_i, i = 0, 1, \ldots,$ are determined by

$$c_i = \frac{\langle f(x), \phi_i(x)\rangle}{\langle \phi_i(x), \phi_i(x)\rangle} \qquad i = 0, 1, \ldots. \tag{2.9}$$

Proof. In (2.8), if we take the inner product with $\phi_i(x)$ for both sides, we have

$$\langle f(x), \phi_i(x)\rangle = \sum_{j=0}^{\infty} c_j\langle \phi_j(x), \phi_i(x)\rangle. \tag{2.10}$$

From (2.6) and (2.7), we have

$$\langle f(x), \phi_i(x)\rangle = c_i\langle \phi_i(x), \phi_i(x)\rangle. \tag{2.11}$$

Therefore, we have (2.9). □

---

[1] For complex functions $f(x)$ and $g(x)$, their inner product is

$$\langle f(x), g(x)\rangle = \int_a^b f(x)g^*(x)\,dx.$$

If $\{\phi_i\}_{i=0}^{\infty}$ is orthonormal, then (2.8) can be simplified to

$$c_i = \langle f(x), \phi_i(x) \rangle \qquad i = 0, 1, \ldots. \tag{2.12}$$

The above theorem describes how to compute the coefficients, given that we know the function can be represented as a linear combination of an orthogonal function family. How can we find the coefficients of the linear combination? The general theory is derived from the theory of completeness and convergence [96], which is beyond the scope of this book.

2.1.2 Fourier Series

To show that the trigonometric function family is orthogonal, we need the following lemma that should be easy for you to prove if you have understood the material so far.

Lemma 2.5. *For an integer n,*

$$\int_{-\pi}^{\pi} \cos nx \, dx = \begin{cases} 2\pi & \text{if } n = 0 \\ 0 & \text{otherwise} \end{cases} \tag{2.13}$$

and

$$\int_{-\pi}^{\pi} \sin nx \, dx = 0. \tag{2.14}$$

The next theorem states that the trigonometric function family is orthogonal in $[-\pi, \pi]$.

Theorem 2.6. *For integers $n, m \geq 0$,*

$$\int_{-\pi}^{\pi} \cos nx \cos mx \, dx = \begin{cases} 2\pi & m = n = 0 \\ \pi & m = n \neq 0 \\ 0 & m \neq n \end{cases} \tag{2.15}$$

$$\int_{-\pi}^{\pi} \sin nx \sin mx \, dx = \begin{cases} \pi & m = n \neq 0 \\ 0 & \text{otherwise} \end{cases} \tag{2.16}$$

and

$$\int_{-\pi}^{\pi} \cos mx \sin nx \, dx = 0. \tag{2.17}$$

Proof. Trigonometry tells us that

$$\cos x \cos y = \tfrac{1}{2}\big(\cos(x+y) + \cos(x-y)\big),$$
$$\sin x \sin y = \tfrac{1}{2}\big(\cos(x-y) - \cos(x+y)\big),$$
$$\sin x \cos y = \tfrac{1}{2}\big(\cos(x+y) + \sin(x-y)\big).$$

For $m \neq n$, from Lemma 2.5 we have

$$\int_{-\pi}^{\pi} \cos nx \cos mx \, dx = \int_{-\pi}^{\pi} \tfrac{1}{2}\big(\cos(n+m)x + \cos(n-m)x\big)\, dx$$

$$= \begin{cases} \int_{-\pi}^{\pi} \tfrac{1}{2}\big(\cos 0 + \cos 0\big)\, dx = 2\pi & m = n = 0 \\ \int_{-\pi}^{\pi} \tfrac{1}{2}\big(\cos 2nx + \cos 0\big)\, dx = \pi & m = n \neq 0 \\ \int_{-\pi}^{\pi} \tfrac{1}{2}\big(\cos(n+m)x + \cos(n-m)x\big)\, dx = 0 & m \neq n \end{cases}$$

and similarly for other cases. □

If we define a function family $\{\phi_k\}_{k=0}^{\infty}$ on the interval $[-\pi, \pi]$ as follows,

$$\begin{cases} \phi_0(x) = 1 \\ \phi_{2i}(x) = \cos(ix) \\ \phi_{2i-1}(x) = \sin(ix), \end{cases} \tag{2.18}$$

then $\{\phi_k\}_{k=0}^{\infty}$ is orthogonal.

Directly applying (2.9), we know that if a function on interval $[-\pi, \pi]$ can be represented by the orthogonal function family above, the coefficients can be computed.

Similarly, for a function defined on $[-T/2, T/2]$, by replacing x with $\frac{\pi x}{T/2}$, we can construct the following orthogonal function family

$$\begin{cases} \phi_0(x) = 1 \\ \phi_{2i}(x) = \cos\left(\frac{2\pi i x}{T}\right) \\ \phi_{2i-1}(x) = \sin\left(\frac{2\pi i x}{T}\right). \end{cases} \tag{2.19}$$

Let $\tilde{f}(x)$ be a function defined on $[-T/2, T/2]$, we can extend $\tilde{f}(x)$ to a periodic function $f(x)$ that is defined on $(-\infty, \infty)$ with a period T. Notice that the functions in (2.19) are all periodic with a period T. If $\tilde{f}(x)$ defined on $[-T/2, T/2]$ can be represented as a linear combination of functions in (2.19), then $f(x)$ on $(-\infty, \infty)$ can also be represented as a linear combination of functions in (2.19). Therefore, we have the following theorem.

Theorem 2.7. *If a periodic function $f(x)$ with a period T defined on $(-\infty, \infty)$ can be represented as follows:*

$$f(x) = a_0 + \sum_{i=1}^{\infty}\left[a_i \cos\left(\frac{2\pi i x}{T}\right) + b_i \sin\left(\frac{2\pi i x}{T}\right)\right]. \tag{2.20}$$

Then

$$
\begin{aligned}
a_0 &= \frac{1}{T}\int_{-T/2}^{T/2} f(x)\,dx \\
a_i &= \frac{2}{T}\int_{-T/2}^{T/2} f(x)\cos\left(\frac{2\pi i x}{T}\right)dx \\
b_i &= \frac{2}{T}\int_{-T/2}^{T/2} f(x)\sin\left(\frac{2\pi i x}{T}\right)dx.
\end{aligned}
\tag{2.21}
$$

Proof. From Theorem 2.4. □

The above representation is called the *infinite Fourier series representation* of $f(x)$. For almost every periodic function in practice, an infinite Fourier series representation exists.

Sometime it is more convenient to present the trigonometric functions as complex exponential functions using the Euler relation. Let $j = \sqrt{-1}$ denote the basic imagine unit. The Euler relation is as follows:

$$
\begin{aligned}
\cos x + j\sin x &= e^{jx} \\
\cos x &= \frac{e^{jx} + e^{-jx}}{2} \\
\sin x &= \frac{e^{jx} - e^{-jx}}{2}.
\end{aligned}
\tag{2.22}
$$

Let

$$
\phi_i(x) = e^{j2\pi ix/T}, \quad i \in \text{integer}; \tag{2.23}
$$

the Fourier series complex representation of $f(x)$ is

$$
f(x) = \sum_{i=-\infty}^{\infty} c_i e^{j2\pi ix/T}, \tag{2.24}
$$

where

$$
c_i = \frac{1}{T}\int_T f(x)e^{-j2\pi ix/T}\,dx. \tag{2.25}
$$

2.1.3 Fourier Transform

The Fourier series is defined only for periodic functions. For general functions, we can still find their representations in the frequency domain. This requires the *Fourier Transform*.

The formal definition of Fourier transform is stated in the following theorem.

Theorem 2.8 (Fourier Transform). *Given a function $f(x)$ of a real variable x, if*

$$\int_{-\infty}^{\infty} |f(x)|\,dx < \infty, \tag{2.26}$$

then the Fourier transform (FT) of $f(x)$ exists, and it is

$$F(\omega) = \frac{1}{\sqrt{2\pi}} \int_{-\infty}^{\infty} f(x)e^{-j\omega x}\,dx. \tag{2.27}$$

The inverse Fourier transform (IFT) of $F(\omega)$ gives $f(x)$:

$$f(x) = \frac{1}{\sqrt{2\pi}} \int_{-\infty}^{\infty} F(\omega)e^{j\omega x}\,d\omega. \tag{2.28}$$

Since $f(t)$ and $F(\omega)$ are a Fourier transform pair, we denote this as

$$F(\omega) = \mathcal{F}[f(x)]$$

$$f(t) = \mathcal{F}^{-1}[F(\omega)].$$

We will not discuss the FT further here. Instead, we will discuss the Discrete Fourier Transform (DFT) for time series in the next section. Conveniently, many properties of the FT are similar to the corresponding properties of the DFT.

2.1.4 Discrete Fourier Transform

Both the Fourier series and Fourier Transform deal with functions. In time series analysis, we are given a finite sequence of values observed at discrete times. We cannot therefore apply the FT directly to time series. In this section, we discuss the DFT for time series.

The DFT will map a sequence in the time domain into another sequence in the frequency domain. Here is the definition of the DFT.

Definition 2.9 (Discrete Fourier Transform). *Given a time sequence $\mathbf{x} = (x(0), x(1), \ldots, x(N-1))$, its Discrete Fourier Transform (DFT) is*

$$\mathbf{X} = \mathrm{DFT}(\mathbf{x}) = (X(0), X(1), \ldots, X(N-1)),$$

where

$$X(F) = \frac{1}{\sqrt{N}} \sum_{i=0}^{N-1} x(i)e^{-j2\pi Fi/N} \qquad F = 0, 1, \ldots, N-1 \tag{2.29}$$

The Inverse Discrete Fourier Transform (IDFT) of $\mathbf{X}$, $\mathbf{x} = \mathrm{IDFT}(\mathbf{X})$, is given by

$$x(i) = \frac{1}{\sqrt{N}} \sum_{F=0}^{N-1} X(F)e^{j2\pi Fi/N} \qquad i = 0, 1, \ldots, N-1. \tag{2.30}$$

Note that **X** and **x** are of the same size. The DFT of a time series is another time series. We will see that many operations on the time series and the time series we get from the DFT are closely related.

We sometime denote the above relation as

$$X(F) = \mathrm{DFT}(x(i)). \tag{2.31}$$

If we set

$$W_N = e^{j2\pi/N},$$

the transforms can be written as

$$X(F) = \frac{1}{\sqrt{N}} \sum_{i=0}^{N-1} x(i) W_N^{-Fi} \qquad F = 0, 1, \ldots, N-1, \tag{2.32}$$

and

$$x(i) = \frac{1}{\sqrt{N}} \sum_{F=0}^{N-1} X(F) W_N^{Fi} \qquad i = 0, 1, \ldots, N-1. \tag{2.33}$$

We can also write the DFT and the IDFT in matrix form. Let

$$\mathbf{W} = \begin{pmatrix} W_N^{-0\cdot 0} & W_N^{-0\cdot 1} & \cdots & W_N^{-0\cdot (N-1)} \\ W_N^{-1\cdot 0} & W_N^{-1\cdot 1} & \cdots & W_N^{-1\cdot (N-1)} \\ \vdots & \vdots & \ddots & \vdots \\ W_N^{-(N-1)\cdot 0} & W_N^{-(N-1)\cdot 1} & \cdots & W_N^{-(N-1)\cdot (N-1)} \end{pmatrix} \tag{2.34}$$

and

$$\overline{\mathbf{W}} = \begin{pmatrix} W_N^{0\cdot 0} & W_N^{0\cdot 1} & \cdots & W_N^{0\cdot (N-1)} \\ W_N^{1\cdot 0} & W_N^{1\cdot 1} & \cdots & W_N^{1\cdot (N-1)} \\ \vdots & \vdots & \ddots & \vdots \\ W_N^{(N-1)\cdot 0} & W_N^{(N-1)\cdot 1} & \cdots & W_N^{(N-1)\cdot (N-1)} \end{pmatrix}; \tag{2.35}$$

we have

$$\mathbf{X} = \mathrm{DFT}(\mathbf{x}) = \mathbf{x}\mathbf{W} \tag{2.36}$$

$$\mathbf{x} = \mathrm{IDFT}(\mathbf{X}) = \mathbf{x}\overline{\mathbf{W}}. \tag{2.37}$$

If the time series $X(F)$ and $x(i)$ are complex, their real and imaginary parts of the real time series are

$$x(i) = x_R(i) + jx_I(i),$$

$$X(F) = X_R(F) + jX_I(F).$$

To compute the DFT, we use the following relation:

$$X(F)=\frac{1}{\sqrt{N}}\sum_{i=0}^{N-1}\left(x_R(i)+jx_I(i)\right)\left[\cos\left(\frac{2\pi Fi}{N}\right)-i\sin\left(\frac{2\pi Fi}{N}\right)\right] \quad (2.38)$$

where $F = 0, 1, \ldots, N-1$. Therefore,

$$X_R(F)=\frac{1}{\sqrt{N}}\sum_{i=0}^{N-1}\left[x_R(i)\cos\left(\frac{2\pi Fi}{N}\right)+x_I(i)\sin\left(\frac{2\pi Fi}{N}\right)\right] \quad (2.39)$$

and

$$X_I(F)=\frac{1}{\sqrt{N}}\sum_{i=0}^{N-1}\left[x_I(i)\cos\left(\frac{2\pi Fi}{N}\right)-x_R(i)\sin\left(\frac{2\pi Fi}{N}\right)\right] \quad (2.40)$$

where $F = 0, 1, \ldots, N-1$. Similarly for the IDFT, we have

$$x_R(i)=\frac{1}{\sqrt{N}}\sum_{i=0}^{N-1}\left[X_R(F)\cos\left(\frac{2\pi Fi}{N}\right)-X_I(F)\sin\left(\frac{2\pi Fi}{N}\right)\right] \quad (2.41)$$

and

$$x_I(i)=\frac{1}{\sqrt{N}}\sum_{i=0}^{N-1}\left[X_I(F)\cos\left(\frac{2\pi Fi}{N}\right)+X_R(F)\cos\left(\frac{2\pi Fi}{N}\right)\right] \quad (2.42)$$

where $i = 0, 1, \ldots, N-1$.

The time series $\mathbf{x} = \big(x(0), x(1), \ldots, x(N-1)\big)$ can be viewed as samples from a function $f(x)$ on the interval $[0, T]$ such that $x_i = f(\frac{i}{N}T)$. Although the Discrete Fourier Transform and Inverse Discrete Fourier Transform are defined only over the finite interval $[0, N-1]$, in many cases it is convenient to imagine that the time series can be extended outside the interval $[0, N-1]$ by repeating the values in $[0, N-1]$ periodically. When we compute the DFT based on (2.29), if we extend the variable F outside the interval $[0, N-1]$, we will actually get a periodical time series. A similar result holds for the IDFT. This periodic property of the DFT and the IDFT is stated in the following theorem.

Theorem 2.10 (Periodicity). *The Discrete Fourier Transform and Inverse Discrete Fourier Transform are periodic with period N:*

$$(a) \quad X(F+N) = X(F) \quad (2.43)$$

$$(b) \quad x(i+N) = x(i) \quad (2.44)$$

$$(c) \quad X(N-F) = X(-F) \quad (2.45)$$

$$(d) \quad x(N-i) = x(-i). \quad (2.46)$$

Proof. (a)

$$\begin{aligned} X(F+N) &= \frac{1}{\sqrt{N}} \sum_{i=0}^{N-1} x(i) e^{-j2\pi(F+N)i/N} \\ &= \frac{1}{\sqrt{N}} \sum_{i=0}^{N-1} x(i) e^{-j2\pi Fi/N} e^{-j2\pi i} \\ &= \frac{1}{\sqrt{N}} \sum_{i=0}^{N-1} x(i) e^{-j2\pi Fi/N} \\ &= X(F). \end{aligned}$$

(b) Similar to (a).

(c)

$$\begin{aligned} X(N-F) &= \frac{1}{\sqrt{N}} \sum_{i=0}^{N-1} x(i) e^{-j2\pi(N-F)i/N} \\ &= \frac{1}{\sqrt{N}} \sum_{i=0}^{N-1} x(i) e^{-j2\pi(-F)i/N} e^{-j2\pi i} \\ &= \frac{1}{\sqrt{N}} \sum_{i=0}^{N-1} x(i) e^{-j2\pi(-F)i/N} \\ &= X(-F). \end{aligned}$$

(d) Similar to (c). □

Theorem 2.11 (Linearity). *The Discrete Fourier Transform and Inverse Discrete Fourier Transform are linear transforms. That is, if*

$$x(i) = ay(i) + bz(i) \tag{2.47}$$

then

$$X(F) = aY(F) + bZ(F). \tag{2.48}$$

Proof. This is obvious from (2.36) and the linearity of matrix multiplication. □

Recall that the transform matrices $\mathbf{W}$ and $\overline{\mathbf{W}}$ are symmetric. This leads to the symmetric properties of the DFT and the IDFT. The following theorems describe the symmetry relation of the DFT and the IDFT.

Theorem 2.12 (Symmetry 1).

$$(a)\ X(-F) = \mathrm{DFT}(x(-i)) \tag{2.49}$$

$$(b)\ X^*(F) = \mathrm{DFT}(x^*(-i)) \tag{2.50}$$

X^* denotes the complex conjugate of X.

Proof. (a)

$$\begin{aligned}
\mathrm{DFT}(x(-i)) &= \mathrm{DFT}(x(N-i)) \\
&= \frac{1}{\sqrt{N}} \sum_{i=0}^{N-1} x(N-i) e^{-j2\pi Fi/N} \\
&= \frac{1}{\sqrt{N}} \sum_{k=1}^{N} x(k) e^{-j2\pi F(N-k)/N} && (k = N - i) \\
&= \frac{1}{\sqrt{N}} \sum_{k=0}^{N-1} x(k) e^{-j2\pi F(N-k)/N} && \text{(periodicity)} \\
&= \frac{1}{\sqrt{N}} \sum_{k=0}^{N-1} x(k) e^{-j2\pi F(-k)/N} \\
&= X(-F).
\end{aligned}$$

(b) Similar to (a) of this proof. □

Theorem 2.13 (Symmetry 2).
(a) If $x(i)$ is even[2]*, then $X(F)$ is even.*

(b) If $x(i)$ is odd[3]*, then $X(F)$ is odd.*

Proof. (a) If $x(i)$ is even, then $x(i) = x(-i)$.
From Theorem 2.12,

$$\mathrm{DFT}(x(-i)) = X(-F). \tag{2.51}$$

Therefore

$$X(F) = \mathrm{DFT}(x(i)) = \mathrm{DFT}(x(-i)) = X(-F). \tag{2.52}$$

So by definition, $X(F)$ is even.
(b) Similar to (a). □

Theorem 2.14 (Symmetry 3).
(a) If $x(i)$ is real, then $X(F) = X^*(-F)$.

(b) If $X(F)$ is real, then $x(i) = x^*(-i)$.

Proof. (a)

$$X(-F) = \frac{1}{\sqrt{N}} \sum_{i=0}^{N-1} x(i) W_N^{Fi}. \tag{2.53}$$

---

[2] A sequence $x(i)$ is even if and only if $x(i) = x(-i)$.
[3] A sequence $x(i)$ is odd if and only if $x(i) = -x(-i)$.

Because $x(i)$ is real, $x^*(i) = x(i)$.

$$\begin{aligned} X^*(-F) &= \frac{1}{\sqrt{N}} \sum_{i=0}^{N-1} x^*(i) W_N^{-Fi} \\ &= \frac{1}{\sqrt{N}} \sum_{i=0}^{N-1} x(i) W_N^{-Fi} \\ &= X(F). \end{aligned}$$

(b) Similar to (a) of this proof. □

The following shifting theorem is very important for the fast computation of the DFT and the IDFT.

Theorem 2.15 (Shifting). *For* $n \in [0, N-1]$, *we have*

$$(a) \quad \mathrm{DFT}\big(x(i-n)\big) = X(F) W_N^{-nF} \tag{2.54}$$

$$(b) \quad \mathrm{IDFT}\big(X(F-n)\big) = x(i) W_N^{ni}. \tag{2.55}$$

Proof. (a). In (2.29), substituting $i-n$ for i, we have

$$\begin{aligned} \mathrm{DFT}\big(x(i-n)\big) &= \frac{1}{\sqrt{N}} \sum_{i=0}^{N-1} x(i-n) W_N^{-Fi} \\ &= \frac{1}{\sqrt{N}} \sum_{k=-n}^{N-1-n} x(k) W_N^{-F(k+n)} \qquad (k = i-n) \\ &= \frac{1}{\sqrt{N}} \sum_{k=0}^{N-1} x(k) W_N^{-Fk} W_N^{-Fn} \\ &= X(F) W_N^{-Fn}. \end{aligned}$$

(b). Similar to (a) of this proof. □

Convolution and product are two important operations between pair of time series. Their definitions are as follows:

Definition 2.16 (Convolution). *The convolution of two time series*

$$\mathbf{x} = \big(x(0), x(1), \ldots, x(N-1)\big) \quad \textit{and} \quad \mathbf{y} = \big(y(0), y(1), \ldots, y(N-1)\big)$$

is another time series $\mathbf{z} = \big(z(0), z(1), \ldots, z(N-1)\big)$, *where*

$$z(i) = \sum_{n=0}^{N-1} x(n) y(i-n), \qquad i = 0, 1, \ldots, N-1, \tag{2.56}$$

denoted as

$$\mathbf{z} = \mathbf{x} \otimes \mathbf{y}. \tag{2.57}$$

Note that in (2.56), $y(i-n) = y(N+i-n)$ for $i-n<0$. Therefore, such a convolution is also known as a circular convolution. Actually, for convolution the two time series need not be of the same length. For time series of different lengths, we can just pad the shorter time series with zeroes to make it as long as the longer one.

Definition 2.17 (Product). *The product of two time series*

$$\mathbf{x} = \big(x(0), x(1), \ldots, x(N-1)\big) \quad \textit{and} \quad \mathbf{y} = \big(y(0), y(1), \ldots, y(N-1)\big)$$

is another time series $\mathbf{z} = \big(z(0), z(1), \ldots, z(N-1)\big)$, *where*

$$z(i) = x(i)y(i), \qquad i = 0, 1, \ldots, N-1, \tag{2.58}$$

denoted as

$$\mathbf{z} = \mathbf{x} * \mathbf{y}. \tag{2.59}$$

It turns out that convolution and product are symmetric through the Fourier Transform.

Theorem 2.18 (Convolution).

$$\text{DFT}\big(\mathbf{x} \otimes \mathbf{y}\big) = \sqrt{N}\mathbf{X} * \mathbf{Y}. \tag{2.60}$$

Proof. DFT($\mathbf{x}\otimes\mathbf{y}$) is an N-dimensional vector corresponding to $F = 0, 1, \ldots N-1$. The Fth element of DFT($\mathbf{x} \otimes \mathbf{y}$) is

$$\text{DFT}(\mathbf{x} \otimes \mathbf{y})[F] = \frac{1}{\sqrt{N}} \sum_{i=0}^{N-1} \Big[\sum_{n=0}^{N-1} x(n)y(i-n)\Big] W_N^{-Fi}.$$

Similarly, $\mathbf{X} * \mathbf{Y}$ is an N dimensional vector corresponding to $F = 0, 1, \ldots N-1$. The Fth element of $\mathbf{X} * \mathbf{Y}$ is

$$\begin{aligned}
\mathbf{X} * \mathbf{Y}[F] &= \Big[\frac{1}{\sqrt{N}} \sum_{n=0}^{N-1} x(n) W_N^{-Fn}\Big]\Big[\frac{1}{\sqrt{N}} \sum_{k=0}^{N-1} y(k) W_N^{-Fk}\Big] \\
&= \frac{1}{N} \sum_{n=0}^{N-1} \Big[\sum_{k=0}^{N-1} x(n)y(k) W_N^{-F(n+k)}\Big] \\
&= \frac{1}{N} \sum_{n=0}^{N-1} \Big[\sum_{i=n}^{N+n-1} x(n)y(i-n) W_N^{-Fi}\Big] \qquad (i = k+n) \\
&= \frac{1}{N} \sum_{n=0}^{N-1} \Big[\sum_{i=0}^{N-1} x(n)y(i-n) W_N^{-Fi}\Big] \qquad (\text{Periodicity : offset of } n) \\
&= \frac{1}{N} \sum_{i=0}^{N-1} \Big[\sum_{n=0}^{N-1} x(n)y(i-n)\Big] W_N^{-Fi}.
\end{aligned}$$

Therefore for each value of F, the two calculations are a factor of $\sqrt{N}$ apart. Hence

$$\text{DFT}\big(\mathbf{x} \otimes \mathbf{y}\big) = \sqrt{N}\mathbf{X} * \mathbf{Y}.$$

□

Theorem 2.19 (Product).

$$\mathrm{DFT}\big(\mathbf{x} * \mathbf{y}\big) = \sqrt{N}\mathbf{X} \otimes \mathbf{Y} \tag{2.61}$$

Proof. Similar to the proof of Theorem 2.18. □

The time complexity to compute the convolution of time series of length n is $O(n^2)$, while the complexity of the product is $O(n)$. The above two theorems connect convolution with product. Therefore, instead of computing the convolution directly, we can perform the DFT on the time series, compute their product and perform the IDFT on the product. This will take time $O(n) + T(n)$, where $T(n)$ is the time complexity of the DFT. In the next section we will show a Fast Fourier Transform (FFT) that takes time $O(n \log n)$, therefore the time complexity of convolution can be reduced to $O(n \log n)$ thanks to the FFT.

Inner product is another important operations between pair of time series.

Definition 2.20 (Inner Product). *The inner product of two time series*

$$\mathbf{x} = \big(x(0), x(1), \ldots, x(N-1)\big) \quad \textit{and} \quad \mathbf{y} = \big(y(0), y(1), \ldots, y(N-1)\big)$$

is

$$\langle \mathbf{x}, \mathbf{y} \rangle = \sum_{i=0}^{N-1} x(i) y^*(i). \tag{2.62}$$

The inner product between a pair of real time series is just

$$\langle \mathbf{x}, \mathbf{y} \rangle = \sum_{i=0}^{N-1} x(i) y(i). \tag{2.63}$$

The following inner product theorem (also known as a power theorem) says that the DFT preserves the inner product between two time series.

Theorem 2.21 (Inner Product).

$$\langle \mathbf{x}, \mathbf{y} \rangle = \langle \mathbf{X}, \mathbf{Y} \rangle \tag{2.64}$$

Proof. Let

$$\mathbf{z} = \mathbf{x} \otimes \mathbf{y}',$$

where $\mathbf{y}' = \big(y^*(-i)\big)$. From Theorem 2.12,

$$\mathrm{DFT}(\mathbf{y}') = \mathbf{Y}^*.$$

From Definition 2.16,

$$z(0) = \sum_{n=0}^{N-1} x(-n) y'(n) = \sum_{n=0}^{N-1} x(-n) y^*(-n).$$

Therefore,

$$\langle \mathbf{x}, \mathbf{y} \rangle = \sum_{k=0}^{N-1} x(k)y^*(k) = \sum_{m=0}^{N-1} x(-m)y^*(-m) = z(0). \tag{2.65}$$

Also from Theorem 2.18,

$$\begin{aligned} \mathbf{Z} &= \mathrm{DFT}(\mathbf{z}) \\ &= \sqrt{N}\mathrm{DFT}(\mathbf{x}) * \mathrm{DFT}(\mathbf{y}') \\ &= \sqrt{N}\mathbf{X} * \mathbf{Y}^*. \end{aligned}$$

From (2.30),

$$z(0) = \frac{1}{\sqrt{N}} \sum_{F=0}^{N-1} Z(F) = \sum_{F=0}^{N-1} X(F)Y^*(F) = \langle \mathbf{X}, \mathbf{Y} \rangle. \tag{2.66}$$

Comparing (2.65) and (2.66), we have

$$\langle \mathbf{x}, \mathbf{y} \rangle = \langle \mathbf{X}, \mathbf{Y} \rangle.$$ □

You should note that there are three notions of product being used here: product (Definition 2.17), convolution (Definition 2.16) and inner product (Definition 2.20). The results of product and convolution between time series are still time series. The product and convolution are related according to Theorems 2.19 and 2.18. That is, the product of two time series is proportional to the inverse Fourier Transform (IFT) of the convolution of their FTs. Similarly, the convolution of two time series is proportional to the IFT of the product of their FTs. By contrast, the inner product of two time series is a number, which is the sum of the pointwise products of the two time series. Because the FT is orthonormal, the inner product between two time series is the same as the inner product of their Fourier Transforms. Table 2.1 compares the three products.

Table 2.1. Comparison of product, convolution and inner product.

Name	Operation	Alternate computation method
convolution	$\mathbf{x} \otimes \mathbf{y}$	$\sqrt{N}\,\mathrm{IDFT}\big(\mathrm{DFT}(\mathbf{x}) * \mathrm{DFT}(\mathbf{y})\big)$
product	$\mathbf{x} * \mathbf{y}$	$\sqrt{N}\,\mathrm{IDFT}\big(\mathrm{DFT}(\mathbf{x}) \otimes \mathrm{DFT}(\mathbf{y})\big)$
inner product	$\langle \mathbf{x}, \mathbf{y} \rangle$	$\langle \mathrm{DFT}(\mathbf{x}), \mathrm{DFT}(\mathbf{y}) \rangle$

The energy of a real time series is defined as the sum of the squares of each item in the time series.

Definition 2.22. *The energy of a time series* $\mathbf{x} = \big(x(0), x(1), \ldots, x(N-1)\big)$ *is*

$$En(\mathbf{x}) = ||\mathbf{x}||^2 = \sum_{i=0}^{N-1} x^2(i). \tag{2.67}$$

Obviously, we have

$$En(\mathbf{x}) = \langle \mathbf{x}, \mathbf{x} \rangle. \tag{2.68}$$

Finally, we recall the *Rayleigh Energy Theorem (Parseval's Theorem)* which states that the DFT preserves the energy of a time series.

Theorem 2.23 (Rayleigh Energy).

$$||\mathbf{x}||^2 = ||\mathbf{X}||^2. \tag{2.69}$$

Proof. In (2.64), let $\mathbf{y} = \mathbf{x}$, we have $\langle \mathbf{x}, \mathbf{x} \rangle = \langle \mathbf{X}, \mathbf{X} \rangle$. □

A direct consequence of the Rayleigh Energy Theorem is that the DFT preserves the Euclidean distance between time series.

Theorem 2.24 (Euclidean distance). *Let* $\mathbf{X} = \mathrm{DFT}(\mathbf{x}), \mathbf{Y} = \mathrm{DFT}(\mathbf{y})$,

$$||\mathbf{x} - \mathbf{y}||^2 = ||\mathbf{X} - \mathbf{Y}||^2. \tag{2.70}$$

2.1.5 Fast Fourier Transform

The symmetry of the DFT and the IDFT make it possible to compute the DFT efficiently. Cooley and Tukey [23] published a fast algorithm for DFT in 1965. It is known as *Fast Fourier Transform (FFT)*. FFT is one of the most important inventions in computational techniques in the last century. It reduced the computation of the DFT significantly.

From (2.29), we can see that the time complexity of the DFT for a time series of length n is $O(n^2)$. This can be reduced to $O(n \log n)$ using the FFT.

Let $N = 2M$, we have

$$W_N^2 = e^{-j2\pi 2/N} = e^{-j2\pi/M} = W_M, \tag{2.71}$$

$$W_N^M = e^{-j2\pi M/N} = e^{-j\pi} = -1. \tag{2.72}$$

Define $a(i) = x(2i)$ and $b(i) = x(2i+1)$, let their DFTs be $A(F) = \mathrm{DFT}\big(a(i)\big)$ and $B(F) = \mathrm{DFT}\big(b(i)\big)$, we have

$$\begin{aligned} X(F) &= \frac{1}{\sqrt{N}} \sum_{i=0}^{N-1} x(i) W_N^{-Fi} \\ &= \frac{1}{\sqrt{N}} \Big[\sum_{i=0}^{M-1} x(2i) W_N^{-2Fi} + \sum_{i=0}^{M-1} x(2i+1) W_N^{-(2i+1)F} \Big] \\ &= \frac{1}{\sqrt{N}} \Big[\sum_{i=0}^{M-1} x(2i) W_M^{-2Fi} + W_N^{-F} \sum_{i=0}^{M-1} x(2i+1) W_M^{-2Fi} \Big] \\ &= A(F) + W_N^{-F} B(F). \end{aligned} \tag{2.73}$$

If $0 \leq F < M$, then

$$X(F) = A(F) + W_N^{-F} B(F). \tag{2.74}$$

Because $A(F)$ and $B(F)$ have period M, for $0 \leq F < M$, we also have

$$\begin{aligned} X(F+M) &= A(F+M) + W_N^{-(F+M)} B(F+M) \\ &= A(F) + W_N^{-M} W_N^{-F} B(F) \\ &= A(F) - W_N^{-F} B(F). \end{aligned} \tag{2.75}$$

From the above equations, the DFT of a time series $x(i)$ with length N can be computed from the DFT of two time series with length $N/2$: specifically, $a(i)$ and $b(i)$.

Suppose that computing the FFT of a time series of length N takes time $T(N)$. Computing the transform of $x(i)$ requires the transforms of $a(i)$ and $b(i)$, and the product of W_N^{-F} with $B(F)$. Computing $A(F)$ and $B(F)$ takes time $2T(N/2)$, and the product of two time series with size $N/2$ takes time $N/2$. Thus we have the following recursive equation:

$$T(N) = 2T(N/2) + N/2. \tag{2.76}$$

Suppose that $N = 2^a$ for some integer a, solving the above recursive equation gives

$$T(N) = O(N \log N).$$

Therefore the FFT for a time series with size N, where N is a power of 2, can be computed in time $O(N \log N)$. A time series whose size is not a power of 2 can be padded with trailing zeroes before performing the FFT.

2.1.6 Discrete Fourier Transform – The Bottom Line

Often a long time series can be approximated by just a few coefficients in the frequency domain. We can then use the theorems that allow us to compute the inner product and the convolution rapidly in the frequency domain. The Fast Fourier Transform allows us to go from the time to the frequency and then back to the time domain very rapidly. These speedups are predicated on obtaining a good approximation of the original time series from the first few coefficients. Remarkably this occurs very often in practice as we will now see.

Given a time series $\mathbf{x}$ and its approximation $\tilde{\mathbf{x}}$, we can measure the quality of the approximation by the Euclidean distance between them: $d(\mathbf{x}, \tilde{\mathbf{x}}) = ||\mathbf{x} - \tilde{\mathbf{x}}||^2$. If $d(\mathbf{x}, \tilde{\mathbf{x}})$ is close to zero, we know that $\tilde{\mathbf{x}}$ is a good approximation of $\mathbf{x}$, that is, the two time series have the same raw shape. Let $\mathbf{x}_e = \mathbf{x} - \tilde{\mathbf{x}}$ be the time series representing the approximation errors. The better an approximation of time series $\mathbf{x}$, the closer the energy of $\mathbf{x}_e$ is to zero, the closer are the energies of $\mathbf{x}$ and $\tilde{\mathbf{x}}$.

Because the DFT preserves the Euclidean distance between time series, and because for most real time series the first few coefficients contain most of the energy,

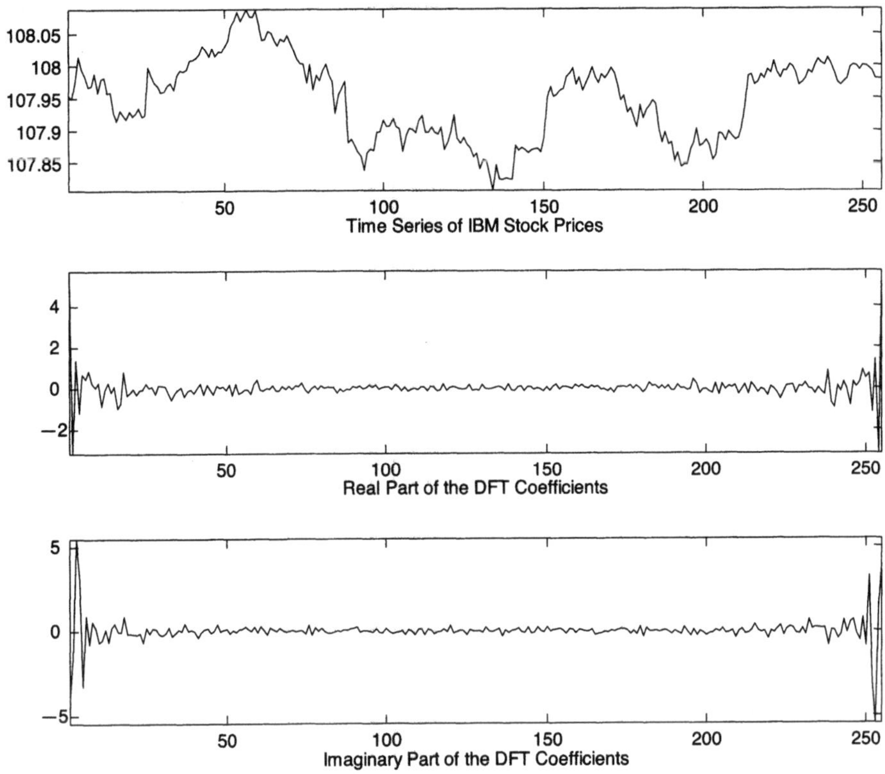

Figure 2.1. IBM stock price time series and its DFT coefficients.

it is reasonable to expect those coefficients to capture the raw shape of the time series [8, 34].

For example, the energy spectrum for the *random walk* series (also known as brown noise or brownian walk), which models stock movements, declines with a power of 2 with increasing coefficients. Figure 2.1 shows a time series of IBM stock prices from 2001 to 2002 and its DFT coefficients. From Theorem 2.14, we know that for a real time series, its k-th DFT coefficient from the beginning are the conjugates of its k-th coefficient from the end. This is verified in the figure. We can also observe that the energy of the time series is concentrated in the first few DFT coefficients (and also the last few coefficients by symmetry).

Birkhoff's theory [84] claims that many interesting signals, such as musical scores and other works of art, consist of *pink noise*, whose energy is concentrated in the first few frequencies (but not as few as in the random walk). On the other hand, *black noise* in the energy spectrum is concentrated even more strongly in the early coefficients than with *brown noise*. Figure 2.2 shows another time series of the ocean level, which is an example of black noise. Its DFT coefficients are also

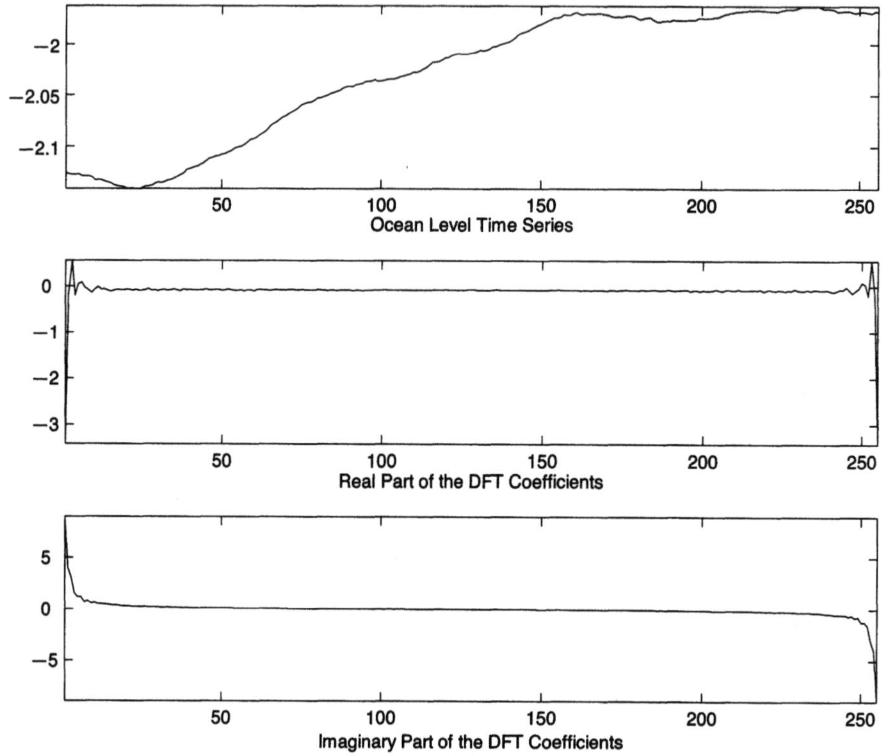

Figure 2.2. Ocean level time series and its DFT coefficients. Data Source: UCR Time Series Data Achieve [56].

shown in the figure. We can see that the energy for this type of time series is more concentrated than the brown noise.

Another type of time series is *white noise*, where each value is completely independent of its neighbors. White noise series has the same energy in every frequency, which implies that all the frequencies are equally important. For pure white noise, there is no way to find a small subset of DFT coefficients that capture most energy of the time series. We will discuss a random projection method as a data reduction technique for time series having large coefficients at all frequencies in Section 2.5.

Data Reduction based on the DFT works by retaining only the first few DFT coefficients of a time series as a concise representation of the time series. For time series modeled by pink, brown and black noises, such a representation will capture most energy of the time series. Note that the symmetry of DFT coefficients for real time series means that the energy contained in the last few DFT coefficients are also used implicitly.

The time series reconstructed from these few DFT coefficients is the DFT approximation of the original time series. Figure 2.3 shows the DFT approximation of the IBM stock price time series. We can see that as we use more and more DFT coefficients, the DFT approximation gets better. But even with only a few DFT coefficients, the major structure of the time series is captured.

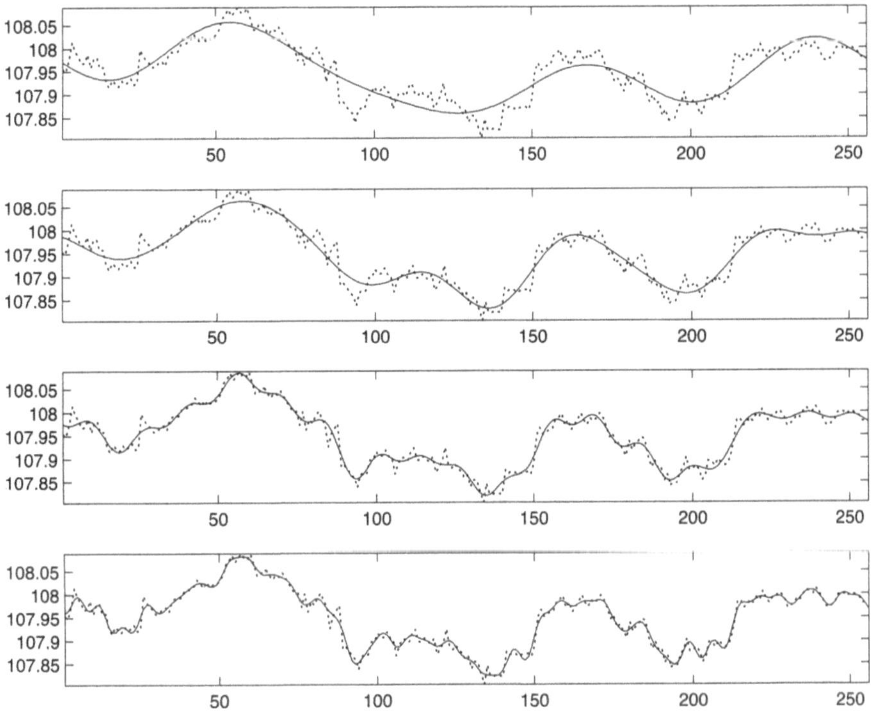

Figure 2.3. Approximation of IBM stock price time series with a DFT. From top to bottom, the time series is approximated by 10, 20, 40 and 80 DFT coefficients, respectively.

In Figure 2.4 we show the approximation of the ocean level time series with the DFT coefficients. We can see that for black noise, fewer DFT coefficients than for brown noise can approximate the time series with high precision.

2.2 Wavelet Transform

Wavelet Analysis follows a similar path to Fourier Analysis. Wavelet Analysis has proved particularly useful for time series that vary significantly over time.

2.2.1 From Fourier Analysis to Wavelet Analysis

The Fourier transform (FT) is ideal for analyzing periodic time series, since the basis functions for Fourier approximation are themselves periodic. The support of the

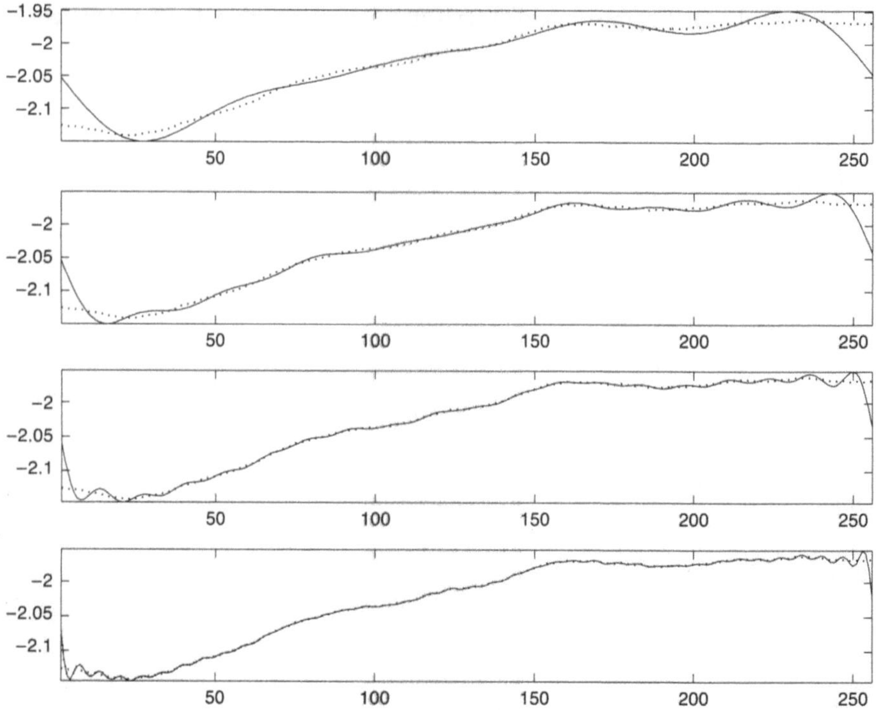

Figure 2.4. Approximation of ocean time series with a DFT. From top to bottom, the time series is approximated by 10, 20, 40 and 80 DFT coefficients respectively.

basis Fourier vectors has the same length as the time series. As a consequence, the sinusoids in FT are very well localized in frequency, but they are not localized in time. This makes it difficult to model spikes economically. We can see this in the following example of an electrocardiogram (ECG) time series.

An electrocardiogram (ECG) time series is an electrical recording of the heart and is used in the investigation of heart disease. An ECG time series is characterized by the spikes corresponding to heartbeats. Figure 2.5 shows an example of an ECG time series and its DFT coefficients. It is impossible to tell when the spikes occur from the DFT coefficients. We can also see that the energy in the frequency domain spread over a relatively large number of DFT coefficients. As a result, the time series approximation using the first few Discrete Fourier Transform coefficients do not give a satisfactory approximation of the time series, especially around the spikes in the original time series. This is demonstrated by Figure 2.6, which shows the approximation of the ECG time series with various numbers of DFT coefficients.

To overcome the above drawback of Fourier Analysis, the *Short Time Fourier Transform (STFT)* [72], also known as *Windowed Fourier Transform (WFT)*, has been used. To represent the frequency behavior of a time series locally in time, the time series is analyzed by functions that are localized both in time and frequency. The

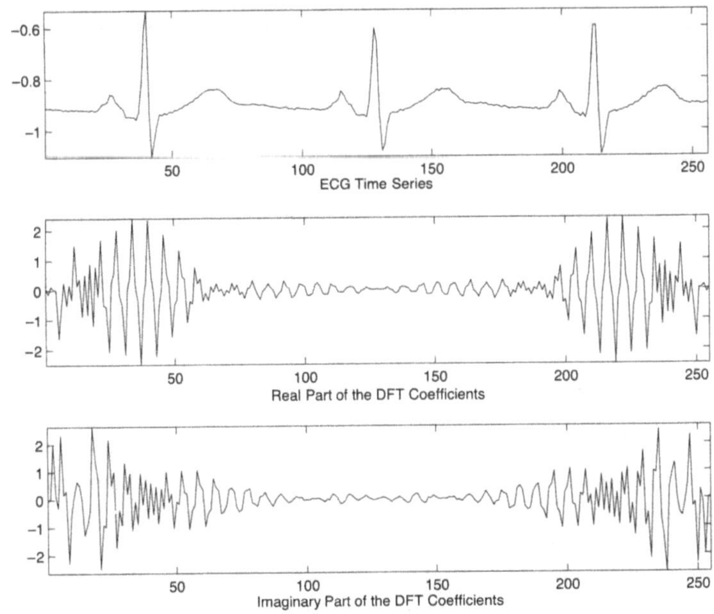

Figure 2.5. An electrocardiogram time series and its DFT coefficients.

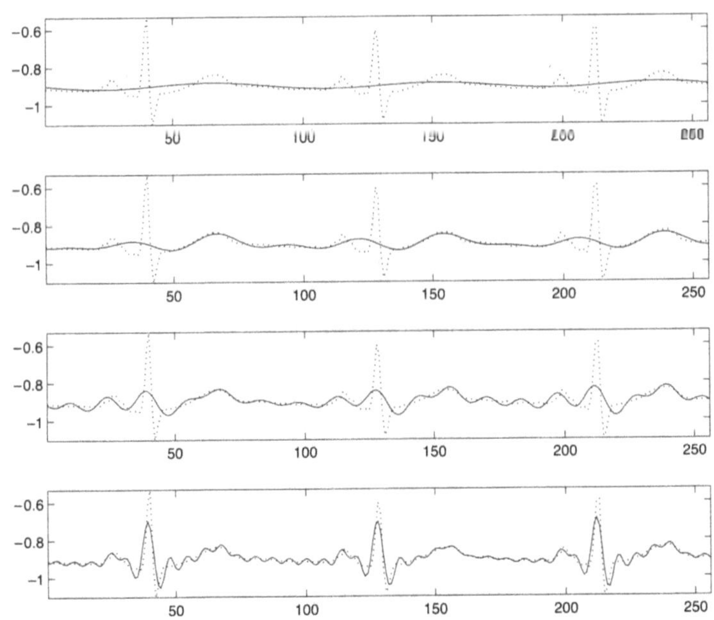

Figure 2.6. Approximations of ECG time series with DFT. From top to bottom, the time series is approximated by 10, 20, 40 and 80 DFT coefficients respectively.

STFT replaces the FT's sinusoidal wave by the product of a sinusoid and a window that is localized in time. Sliding windows of fixed size are imposed on the time series, and the STFT computes the FT in each window. This is a compromise between the time domain and frequency domain analysis of time series. The drawback of the STFT is that the sliding window size has to be fixed and thus the STFT might not provide enough information about the time series as a whole.

The Wavelet Transform (WT) is, in an operational sense, a generalization of the STFT. In the WT, variable-sized windows replace the fixed window size in STFT. Also the sinusoidal waves in FT are replaced by a family of functions called *wavelets*. This results in a Time/Scale Domain analysis of the time series. Scale defines a subsequence of time series under consideration. The scale information is closely related to the frequency information. We will discuss more details of the wavelet analysis in the remaining section.

In Figure 2.7, we compare the four views of a time series: Time Domain analysis, Frequency Domain analysis by the FT, Time/Frequency Domain analysis by the STFT and Time/Scale Domain analysis by the WT. In the Wavelet Transform, higher scales correspond to lower frequencies. We can see that for the Wavelet Transform, the time resolution is better for higher frequencies (smaller scales). By comparison, for the Short Time Fourier Transform the frequency and time resolution are independent.

2.2.2 Haar Wavelet

Let us start with the simplest wavelet, the Haar Wavelet. The Haar Wavelet is based on the step function.

Definition 2.25 (Step function). *A step function is*

$$\chi_{[a,b)}(x) = \begin{cases} 1 & \text{if } a \le x < b, \\ 0 & \text{otherwise.} \end{cases} \tag{2.77}$$

Definition 2.26 (Haar scaling function family). *Let*

$$\phi(x) = \chi_{[0,1)}(x)$$

and

$$\phi_{j,k}(x) = 2^{j/2}\phi(2^j x - k) \qquad j,k \in \mathbf{Z}; \tag{2.78}$$

the collection $\{\phi_{j,k}(x)\}_{j,k\in\mathbf{Z}}$ *is called the system of Haar scaling function family on* $\mathbf{R}$.

Figure 2.8 shows some of the Haar scaling functions on the interval $[0, 1]$. Mathematically, the system of Haar scaling function family, $\phi_{j,k}$, is generated by the Haar scaling function $\phi(x)$ with an integer translation of k (shift) and dyadic dilation (product by the powers of two).

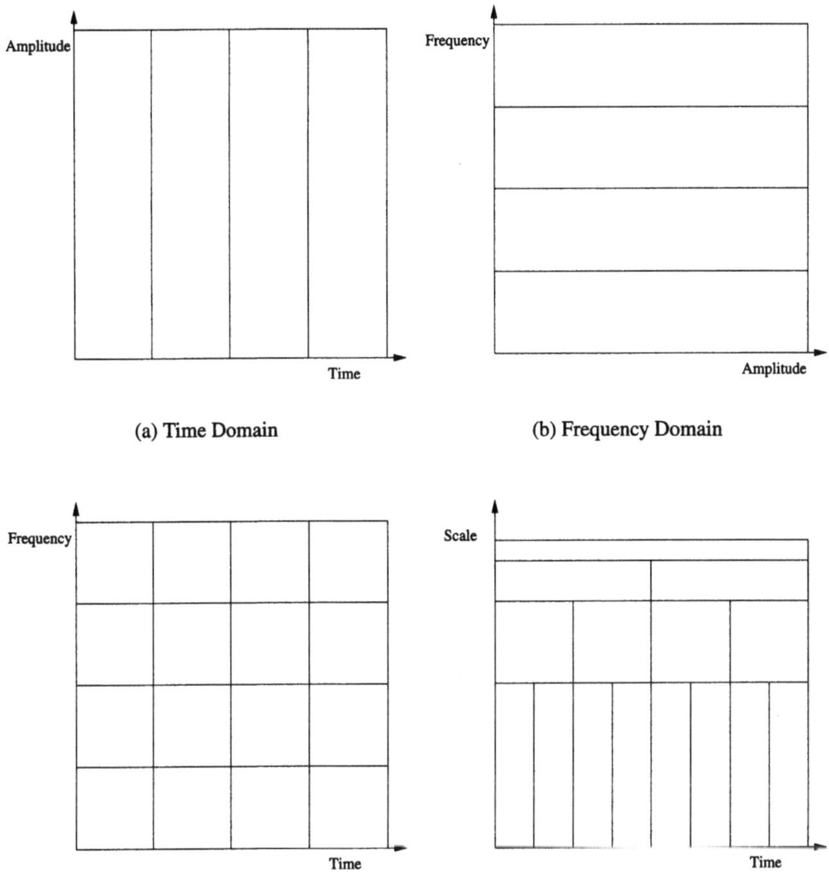

(a) Time Domain (b) Frequency Domain

(c) Time/Frequency Domain (d) Time/Scale Domain

Figure 2.7. Time series analysis in four different domains.

We can see that $\{\phi_{j,k}(x)\}_{k\in\mathbf{Z}}$ for a specific j are a collection of piecewise constant functions. Each piecewise constant function has non-zero support of length 2^{-j}. As j increases, the piecewise constant functions become more and more narrow. Intuitively, any function can be approximated by a piecewise constant function. So, it is not surprising that the Haar scaling function family can approximate any function to any precision.

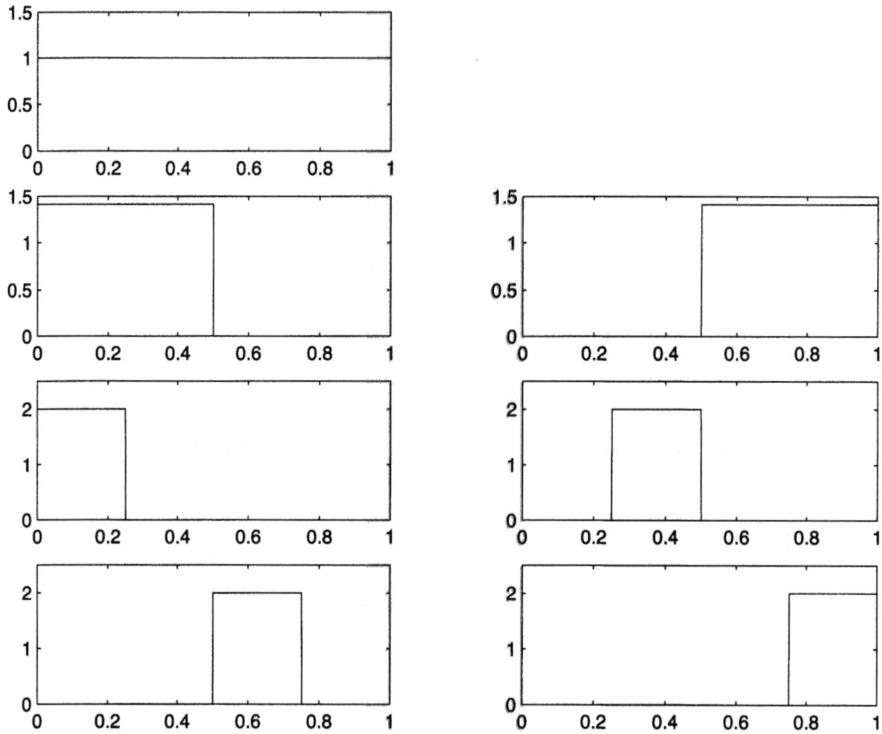

Figure 2.8. Sample Haar scaling functions of Definition 2.26 on the interval [0, 1]. From top to bottom (a) $j = 0, k = 0$; (b) $j = 1, k = 0, 1$; (c) $j = 2, k = 0, 1$; (d) $j = 2, k = 2, 3$.

Definition 2.27 (Haar wavelet function family). *Let*

$$\psi(x) = \chi_{[0,1/2)}(x) - \chi_{[1/2,1)}(x)$$

and

$$\psi_{j,k}(x) = 2^{j/2}\psi(2^j x - k) \qquad j, k \in \mathbf{Z}; \tag{2.79}$$

the collection $\{\psi_{j,k}(x)\}_{j,k\in\mathbf{Z}}$ *is called the system of Haar wavelet function families on* **R***.*

Figure 2.9 shows some of the Haar wavelet functions on the interval [0, 1]. Try to verify that they are orthonormal.

Theorem 2.28. *The Haar wavelet function family on* **R** *is orthonormal.*

2.2.3 Multiresolution Analysis

The Haar function family can approximate functions progressively. This demonstrates the power of *Multiresolution Analysis*, which we will describe in just enough

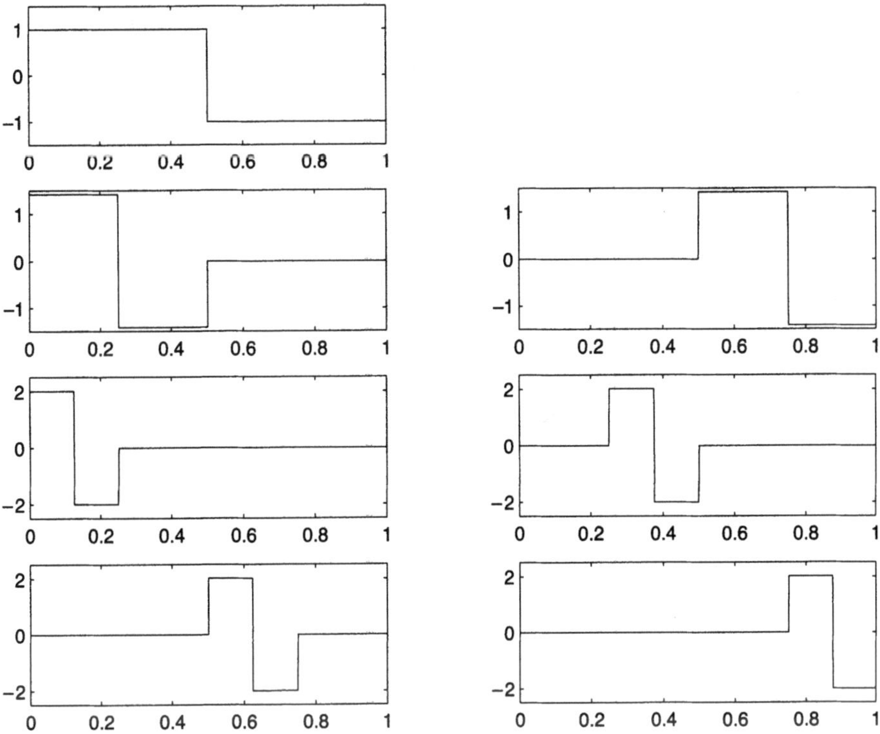

Figure 2.9. Sample Haar wavelet functions of Definition 2.27 on the interval [0, 1]. From top to bottom (a) $j = 0, k = 0$; (b) $j = 1, k = 0, 1$; (c) $j = 2, k = 0, 1$; (d) $j = 2, k = 2, 3$.

detail for our purposes. For a state-of-the-art review of Multiresolution Analysis, please refer to [67, 94].

Definition 2.29 (Multiresolution analysis). *A multiresolution analysis (MRA) on* $\mathbf{R}$ *is a nested sequence of subspaces* $\{V_j\}_{j\in\mathbf{Z}}$ *of functions* L^2 *on* $\mathbf{R}$ *such that*

1. *For all* $j \in \mathbf{Z}$, $V_j \subset V_{j+1}$.
2. $\cap_{j\in\mathbf{Z}} V_j = \{0\}$
3. *For a continuous function* $f(x)$ *on* $\mathbf{R}$, $f(x) \in \cup_{j\in\mathbf{Z}} V_j$.
4. $f(x) \in V_j \Leftrightarrow f(2x) \in V_{j+1}$.
5. $f(x) \in V_0 \Rightarrow f(x-k) \in V_0$.
6. *There exists a function* $\phi(x)$ *on* $\mathbf{R}$, *such that the* $\{\phi(x-k)\}_{k\in\mathbf{Z}}$ *is an orthonormal basis of* V_0.

The first property of multiresolution analysis says that the space V_j is included in space V_{j+1}. Therefore for any function that can be represented by a linear combination of the basis functions of V_j, it can also be represented by a linear combination of the basis functions of V_{j+1}. We can think of the sequence of spaces $V_j, V_{j+1}, V_{j+2}, \ldots$ as the spaces underlying a finer and finer approximation of a function. In the Haar scaling function example, the basis functions of V_j are $\{\phi_{j,k}(x)\}_{k\in\mathbf{Z}}$. Let the projection of $f(x)$ on V_j be $f_j(x)$. $f_j(x)$ approximates $f(x)$ with a piecewise constant function where each piece has length 2^{-j}. We call $f_j(x)$ the approximation function of $f(x)$ at resolution level j. If we add the detailed information at level j, $d_j(x)$, we obtain the approximation of $f(x)$ at level $j+1$. In general, we have

$$f_{j+1}(x) = f_j(x) + d_j(x). \tag{2.80}$$

In other words, the space V_{j+1} can be decomposed into two subspaces V_j and W_j with $f_j(x) \in V_j$ and $d_j(x) \in W_j$. W_j is the detailed space at resolution j and it is orthogonal to V_j. This is denoted by

$$V_{j+1} = V_j \oplus W_j. \tag{2.81}$$

We can expand the above equation as

$$\begin{aligned} V_{j+1} &= V_j \oplus W_j \\ &= V_{j-1} \oplus W_{j-1} \oplus W_j \\ &= \cdots \\ &= V_J \oplus W_J \oplus \cdots \oplus W_{j-1} \oplus W_j \qquad J < j. \end{aligned} \tag{2.82}$$

The above equation says that the approximation space at resolution j can be decomposed into a set of subspaces. The subspaces include the approximation space at resolution J, $J < j$, and all the detailed spaces at resolution between J and j. In general, we can also show that W_j is orthogonal to W_k for all $j \neq k$.

The second property says that the intersection of all the resolution spaces is the 0 space, which includes only the zero function. This can be interpreted as follows. The approximation space will get coarser and coarser as j decreases. When $j \to -\infty$, we cannot have any information of the function in the space. For example, in the Haar wavelet space, if $j \to -\infty$, a function will be approximated by a constant function. That is the coarsest approximation space possible. The requirement that the function must be square integrable gives the zero function.

On the other hand, the third property states that any function can be approximated at a certain resolution. The reason is that as $j \to \infty$, we will have the finest approximation. Any function can thus be approximated if we go up to a certain resolution level.

The fourth property states that all the space must be scaled versions of V_0 and the fifth property states that the space is invariant under translation. These two properties imply that

$$f(x) \in V_0 \Rightarrow f(2^j x - k) \in V_j. \tag{2.83}$$

The function $\phi(x)$ in the last property is called the *scaling function* of the multiresolution analysis. From this function, we will construct the orthonormal basis functions for the nested sequence of subspaces in multiresolution analysis.

2.2.4 Wavelet Transform

Based on this brief introduction to multiresolution analysis, we can introduce the general theory of wavelets. For simplicity, we restrict ourselves here to orthogonal wavelets. The last property of multiresolution analysis says that the scaling function generate the orthonormal basis functions for V_0. In fact, it also generates the orthonormal basis functions for V_j.

Theorem 2.30.

$$\{\phi_{j,k}(x)\}_{k\in\mathbf{Z}} \tag{2.84}$$

is an orthonormal basis on V_j, *where*

$$\phi_{j,k}(x) = 2^{j/2}\phi(2^j x - k). \tag{2.85}$$

Next we examine the relation between adjacent levels of resolution space.

Theorem 2.31. *There exists a coefficient sequence* $\{h_k\}$ *such that*

$$\phi(x) = 2^{1/2}\sum_k h_k \phi(2x - k). \tag{2.86}$$

Proof. We know that $\phi_{1,k}(x)$ are orthonormal basis functions for V_1. Because $V_0 \subset V_1$ and $\phi(x) \in V_0$, we have the following *dilation equation*:

$$\phi(x) = \sum_k h_k \phi_{1,k}(x) = 2^{1/2}\sum_k h_k \phi(2x - k). \tag{2.87}$$

Because $\phi_{1,k}(x)$ are orthonormal basis functions,

$$h_k = \langle \phi_{1,k}(x), \phi(x)\rangle = 2^{1/2}\int_{-\infty}^{\infty} \phi(x)\phi(2x - k)\, dx. \tag{2.88}$$

□

The coefficient sequence $\{h_k\}$ is called the scaling filter.

Symmetric to the scaling function $\phi(x)$, a wavelet function $\psi(x)$ is designed to generate the orthonormal basis function for the detailed space W_j. That is,

$$\{\psi_{j,k}(x)\}_{k\in\mathbf{Z}}$$

is the orthonormal basis functions for the detailed space W_j, where

$$\psi_{j,k}(x) = 2^{j/2}\psi(2^j x - k). \tag{2.89}$$

We will not get into the details of how to design the wavelet function $\psi(x)$ given the scaling function $\phi(x)$. Instead, we will infer some properties of the wavelet function based on the above assertion.

From $W_0 \subset V_1$, $\psi(x) \in W_0$, and the fact that $\phi_{1,k}(x)$ are orthonormal basis functions for V_1, we have the following *wavelet equation*:

$$\psi(x) = \sum_k g_k \phi_{1,k}(x) = 2^{1/2} \sum_k g_k \phi(2x - k). \tag{2.90}$$

Because $\phi_{1,k}(x)$ are orthonormal basis functions,

$$g_k = \langle \phi_{1,k}(x), \psi(x) \rangle = 2^{1/2} \int_{-\infty}^{\infty} \psi(x)\phi(2x-k)\,dx. \tag{2.91}$$

The coefficient sequence $\{g_k\}$ is called the wavelet filter.

The following theorem gives the connection between the scaling filter $\{h_k\}$ and the wavelet filter $\{g_k\}$.

Theorem 2.32. *Given a scaling filter h_k, the wavelet filter g_k is*

$$g_k = (-1)^k h_{1-k}^*. \tag{2.92}$$

The scaling functions at adjacent levels of resolution space are connected by the scaling filter. Similarly, the wavelet filter bridges the wavelet functions of adjacent resolution levels. This is stated by the following theorem.

Theorem 2.33.

$$(a) \quad \phi_{j-1,k}(x) = \sum_l h_{l-2k} \phi_{jl}(x) \tag{2.93}$$

$$(b) \quad \psi_{j-1,k}(x) = \sum_l g_{l-2k} \phi_{jl}(x). \tag{2.94}$$

Proof. (a) Because $V_{j-1} \subset V_j$, basis function $\phi_{j-1,k}$ of V_{j-1} can be represented by $\phi_{j,k}$.

$$\phi_{j-1,k}(x) = \sum_l \langle \phi_{jl}, \phi_{j-1,k} \rangle \phi_{jl}(x), \tag{2.95}$$

and

$$\begin{aligned}
\langle \phi_{jl}, \phi_{j-1,k} \rangle &= \int_{-\infty}^{\infty} 2^{j/2} 2^{(j-1)/2} \phi(2^j x - l)\phi(2^{j-1}x - k)\,dx \\
&= 2^{1/2} \int_{-\infty}^{\infty} \phi(2^j x - l)\phi(2^{j-1}x - k) 2^{j-1}\,dx \\
&= 2^{1/2} \int_{-\infty}^{\infty} \phi(2u + 2k - l)\phi(u)du \qquad (u = 2^{j-1}x - k) \\
&= h_{l-2k} \qquad \text{(from (2.88))}.
\end{aligned}$$

(b) Because $W_{j-1} \subset V_j$, basis function $\psi_{j-1,k}$ of W_{j-1} can be represented by $\phi_{j,k}$.

$$\psi_{j-1,k}(x) = \sum_l \langle \psi_{jl}, \phi_{j-1,k} \rangle \phi_{jl}(x), \tag{2.96}$$

and

$$\begin{aligned}\langle \phi_{jl}, \psi_{j-1,k} \rangle &= \int_{-\infty}^{\infty} 2^{j/2} 2^{(j-1)/2} \phi(2^j x - l) \psi(2^{j-1} x - k)\, dx \\ &= 2^{1/2} \int_{-\infty}^{\infty} \phi(2^j x - l) \psi(2^{j-1} x - k) 2^{j-1}\, dx \\ &= 2^{1/2} \int_{-\infty}^{\infty} \phi(2u + 2k - l) \psi(u) du \qquad (u = 2^{j-1} x - k) \\ &= g_{l-2k} \qquad \text{(from (2.91)).}\end{aligned}$$

□

Because $\{\psi_{j,k}(x)\}_{k \in \mathbf{Z}}$ is the orthonormal basis functions for the detailed space W_j, from (2.82), we have the following theorem.

Theorem 2.34.

$$\{\psi_{j,k}(x)\}_{j,k \in \mathbf{Z}} \tag{2.97}$$

is a wavelet orthonormal basis on **R**.
Given any $J \in \mathbf{Z}$,

$$\{\phi_{J,k}(x)\}_{k \in \mathbf{Z}} \cup \{\psi_{j,k}(x)\}_{j > J, k \in \mathbf{Z}} \tag{2.98}$$

is also an orthonormal basis on **R**.

From Theorem 2.34, any function $f(x) \in L^2(\mathbf{R})$ can be uniquely represented as

$$f(x) = \sum_{j,k \in \mathbf{Z}} \langle f, \psi_{j,k} \rangle \psi_{j,k}(x). \tag{2.99}$$

This is called the Wavelet Transform of function $f(x)$.

From multiresolution analysis, we know that a function $f(x)$ can be approximated by multilevel resolution. Suppose that $f_j(x)$ is the approximation at level j, we have

$$f_j(x) = \sum_k f_{jk} \phi_{jk}(x) \tag{2.100}$$

where

$$f_{jk} = \langle \phi_{jk}, f \rangle = \langle \phi_{jk}, f_j \rangle. \tag{2.101}$$

Let us represent $f_j(x)$ by its approximation and details at level $j - 1$:

$$f_j(x) = f_{j-1}(x) + d_{j-1}(x) = \sum_k f_{jk} \phi_{jk}(t) + \sum_k d_{jk} \psi_{jk}(t). \tag{2.102}$$

The Wavelet Transform of a function $f(x)$ is to compute f_{jk} and d_{jk} from $f(x)$. We have the following theorem for the construction of the Wavelet Transform.

Theorem 2.35.

$$(a)\quad f_{j-1,k} = \sum_l h_{l-2k} f_{jl} \tag{2.103}$$

$$(b)\quad d_{j-1,k} = \sum_l g_{l-2k} f_{jl}. \tag{2.104}$$

Proof. (a) From $f_{j-1} \in V_{j-1}$ we get

$$f_{j-1}(x) = \sum_k f_{j-1,k}\phi_{j-1,k}(x) \tag{2.105}$$

where

$$\begin{aligned} f_{j-1,k} &= \langle \phi_{j-1,k}, f\rangle \\ &= \left\langle \sum_l h_{l-2k}\phi_{jl}, f \right\rangle \\ &= \sum_l h_{l-2k}\langle \phi_{jl}, f\rangle \\ &= \sum_l h_{l-2k} f_{jl}. \end{aligned}$$

(b) Similarly, from $d_{j-1} \in W_{j-1}$ we get

$$d_{j-1}(x) = \sum_k d_{j-1,k}\psi_{j-1,k}(x) \tag{2.106}$$

where

$$\begin{aligned} d_{j-1,k} &= \langle \psi_{j-1,k}, f\rangle \\ &= \langle \sum_l g_{l-2k}\phi_{jl}, f\rangle \\ &= \sum_l g_{l-2k}\langle \phi_{jl}, f\rangle \\ &= \sum_l g_{l-2k} f_{jl}. \end{aligned}$$

□

There are two interesting observations we can make from Theorem 2.35. First, the wavelet coefficients at the coarser level, $f_{j-1,k}$ and $d_{j-1,k}$, can be computed from the wavelet coefficients at its next finer level, f_{jk}. This also reflects the multiresolution analysis nature of the Wavelet Transform. Second, in the above computation, we do not use the scaling function or wavelet function directly. We use only the scaling and wavelet filters. This is the basis for the fast wavelet computation method we will discuss in Section 2.2.5.

The Inverse Wavelet Transform will reconstruct a function at approximation level j from its approximation and detailed information at resolution level $j-1$. The following theorem gives the reconstruction formula.

Theorem 2.36.

$$f_{jk} = \sum_l h_{k-2l} f_{j-1,k} + \sum_l g_{k-2l} d_{j-1,k}. \tag{2.107}$$

Proof. From $V_j = V_{j-1} \oplus W_{j-1}$,

$$\begin{aligned} \phi_{jk}(x) &= \sum_l \langle \phi_{j-1,l}, \phi_{jk} \rangle \phi_{j-1,l}(x) + \sum_l \langle \psi_{j-1,l}, \phi_{jk} \rangle \psi_{j-1,l}(x) \\ &= \sum_l h_{k-2l} \phi_{j-1,l}(x) + \sum_l g_{k-2l} \psi_{j-1,l}(x). \end{aligned} \tag{2.108}$$

Therefore

$$\begin{aligned} f_{jk} &= \langle \phi_{jk}, f \rangle \\ &= \sum_l h_{k-2l} \langle \phi_{j-1,l}, f \rangle + \sum_l g_{k-2l} \langle \psi_{j-1,l}, f \rangle \\ &= \sum_l h_{k-2l} f_{j-1,k} + \sum_l g_{k-2l} d_{j-1,k}. \end{aligned}$$ □

As for the Wavelet Transform, the Inverse Wavelet Transform computation requires only scaling and wavelet filters, instead of the scaling and wavelet function.

2.2.5 Discrete Wavelet Transform

Time series can be seen as the discretization of a function. After a quick review of the Wavelet Transform (WT) for functions, in this section we will discuss Discrete Wavelet Transform (DWT) for time series.

First we give a concrete example of how to compute the Discrete Wavelet Transform for Haar wavelets.

The Haar Wavelet Transform is the fastest to compute and easiest to implement in the wavelet family. Suppose we have a time series of length 8;

$$S = (1, 3, 5, 11, 12, 13, 0, 1).$$

To perform a WT on the above series, we first average the signal pairwise to get the new lower-resolution signal with value

$$(2, 8, 12.5, 0.5).$$

To recover the original time series from the four averaged values, we need to store some detail coefficients, i.e., the pairwise differences

$$(-1, -3, -0.5, -0.5).$$

Obviously, from the pairwise average vector and the pairwise difference vector we can reconstruct the original time series without loss of any information. These

Table 2.2. Haar Wavelet decomposition tree

Resolution	a_1	a_2	a_3	a_4	a_5	a_6	a_7	a_8
3	$\frac{a_1+a_2}{\sqrt{2}}$	$\frac{a_3+a_4}{\sqrt{2}}$	$\frac{a_5+a_6}{\sqrt{2}}$	$\frac{a_7+a_8}{\sqrt{2}}$	$\frac{a_1-a_2}{\sqrt{2}}$	$\frac{a_3-a_4}{\sqrt{2}}$	$\frac{a_5-a_6}{\sqrt{2}}$	$\frac{a_7-a_8}{\sqrt{2}}$
2	$\frac{a_1+a_2+a_3+a_4}{2}$		$\frac{a_5+a_6+a_7+a_8}{2}$		$\frac{(a_1+a_2)-(a_3+a_4)}{2}$		$\frac{(a_5+a_6)-(a_7+a_8)}{2}$	
1	$\frac{a_1+a_2+a_3+a_4+a_5+a_6+a_7+a_8}{2\sqrt{2}}$				$\frac{(a_1+a_2+a_3+a_4)-(a_5+a_6+a_7+a_8)}{2\sqrt{2}}$			

two vectors are normalized with a factor $\sqrt{2}/2$. After decomposing the original time series into a lower resolution version with half of the number of entries and a corresponding set of detail coefficients, we repeat this process recursively on the average to get the full decomposition.

In general, the Haar wavelet decomposition tree is shown in Table 2.2. From the analysis above, we know that the Haar wavelet decomposition tree capture all the information of the original time series. Table 2.3 shows the Haar wavelet decomposition tree. The averages of the highest resolution level and the details of all the resolution levels are the DWT of the time series. In our example, the DWT of $(1, 3, 5, 11, 12, 13, 0, 1)$ is therefore

$$(16.2635, -2.1213, -6.0000, 12.0000, -1.4142, -4.2426, -0.7071, -0.7071).$$

Table 2.3. An example of a Haar Wavelet decomposition

Resolution	Averages				Details			
4	1	3	5	11	12	13	0	1
3	2.8284	11.3137	17.6777	0.7071	-1.4142	-4.2426	-0.7071	-0.7071
2	10.0000		13.0000		-6.0000		12.0000	
1	16.2635				-2.1213			

We can see that for a time series of length n, the time complexity of the Haar Wavelet Transform is $O(n)$.

For the DWT based on other wavelet families, the basic operations are similar to the process of continuous Wavelet Transforms (CWT) in Theorems 2.35 and 2.36. The theory of DWTs is closely related to Quadrature Mirror Filter (QMF) theory. We will not get into the technical details here. The reader should look at the excellent book by Mallat [67]. Here we just give the practical algorithm for computing the DWT.

In the CWT, we have function $f_j(x) \in V_j$ and approximate $f_j(x)$ in the $V_{j-1}, V_{j-2}, \ldots$ progressively so that

$$
\begin{aligned}
f_j(x) &= f_{j-1}(x) + d_{j-1}(x) \\
&= f_{j-2}(x) + d_{j-2}(x) + d_{j-1}(x) \\
&= f_J(x) + d_J(x) + d_{J+1}(x) + \cdots + d_{j-1}(x) \qquad J < j.
\end{aligned} \tag{2.109}
$$

Similarly, suppose that a time series representation of a function $f_j(x)$ is $\mathbf{f}_j$, i.e., $\mathbf{f}_j$ is derived from $f_j(x)$ by sampling in a finite interval. Discrete Wavelet Transform will produce the approximate time series $\mathbf{f}_{j-1}, \mathbf{f}_{j-2}, \ldots, \mathbf{f}_J$ and the detail time series $\mathbf{d}_{j-1}, \mathbf{d}_{j-2}, \ldots, \mathbf{d}_J$ from $\mathbf{f}_j$.

Any wavelet family is defined by four filters: decomposition low-pass filter $\{d_n^L\}$, decomposition high-pass filter $\{d_n^H\}$, reconstruction low-pass filter $\{r_n^L\}$ and reconstruction high-pass filter $\{r_n^H\}$. These filters are closely related to the scaling and wavelet filters in Section 2.2.4. These filters can be seen as time series too. The Discrete Wavelet Transform and Inverse Discrete Wavelet Transform (IDWT) are performed by convolution with filters, downsampling and upsampling.

Definition 2.37 (Downsampling). *The purpose of downsampling is to reduce time series data to the portion with even indices. Given time series* $(x(1), x(2), \ldots, x(n))$, *its downsampling is time series* $(x(2), x(4), \ldots, x(\lfloor n/2 \rfloor))$.

Definition 2.38 (Upsampling). *Upsampling is just to insert a* 0 *after every data point in the time series except the last. Given time series* $(x(1), x(2), \ldots, x(n))$, *its upsampling is the time series* $(x(1), 0, x(2), 0, x(3), \ldots, x(n-1), 0, x(n))$.

The convolution of a time series $\mathbf{f}_j$ with a decomposition low-pass filter $\{d_n^L\}$ yields the approximation of the time series $\hat{\mathbf{f}}_{j-1}$ at a coarser level. Also the convolution of $\mathbf{f}_j$ with a decomposition high-pass filter $\{d_n^H\}$ yields the details of the time series $\hat{\mathbf{d}}_{j-1}$. The downsampling on $\hat{\mathbf{f}}_{j-1}$ gives a new time series $\mathbf{f}_{j-1}$. The down sampling on $\hat{\mathbf{d}}_{j-1}$ gives $\mathbf{d}_{j-1}$. This process is repeated for $\mathbf{f}_{j-1}$ until we reach some resolution level J. $\mathbf{d}_{j-1}, \mathbf{d}_{j-2}, \ldots, \mathbf{d}_J$ and $\mathbf{f}_J$ are the Discrete Wavelet Transform of $\mathbf{f}_j$. Figure 2.10 shows the procedure for the DWT.

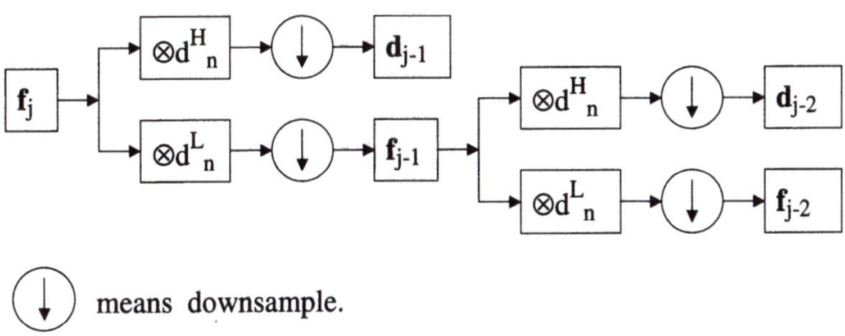

Figure 2.10. Discrete Wavelet Transform with filters and downsampling.

From $\mathbf{d}_{j-1}, \mathbf{d}_{j-2}, \ldots, \mathbf{d}_J$ and $\mathbf{f}_J$ we can also reconstruct the original time series $\mathbf{f}_j$ using the reconstruction filters. This process is called the Inverse Discrete Wavelet

Transform. First we upsample the time series $\mathbf{f}_J$ to get a new time series $\hat{\mathbf{f}}_J$. We also upsample the time series $\mathbf{d}_J$ to get a new time series $\hat{\mathbf{d}}_J$. The sum of the convolution of $\hat{\mathbf{f}}_J$ with $\{r_n^L\}$ and the convolution of $\hat{\mathbf{d}}_J$ with $\{r_n^H\}$ yields $\mathbf{f}_{J+1}$, the approximation at level $J+1$. This process is repeated until we reconstruct $\mathbf{f}_j$. Figure 2.11 shows the procedure of Inverse Discrete Wavelet Transform.

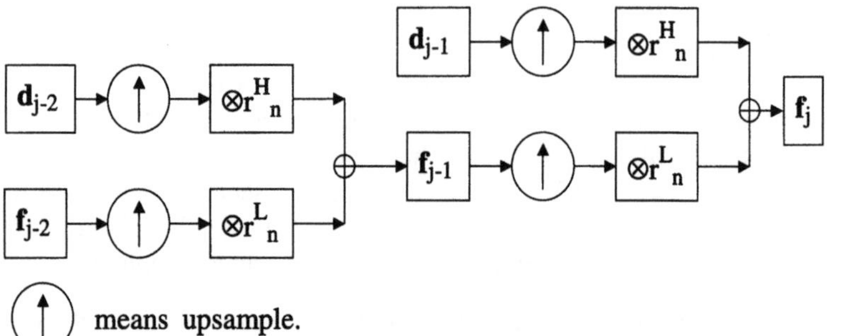

Figure 2.11. Inverse Discrete Wavelet Transform with filters and upsampling.

The above algorithm is also called the *Fast Wavelet Transform algorithm (FWT)* or *pyramid algorithm*. The Wavelet Transform is faster than the FFT. For a time series of length n, the Fourier transform takes times $O(n^2)$ and FFT takes time $O(n \log n)$. In most cases, the time complexity of the Fast Wavelet Transform is $O(n)$.

One of the most popular wavelet families is the Daubechies wavelet family created by Daubechies [30]. The names of the Daubechies family wavelets are written as db_N, where N is the order. The Haar wavelet is actually a Daubechies wavelet with order 1, i.e., db_1. The filters for the Haar wavelet are as follows:

$$
\begin{aligned}
d_n^L &= \left\{\frac{1}{\sqrt{2}}, \frac{1}{\sqrt{2}}\right\}, \qquad d_n^H = \left\{-\frac{1}{\sqrt{2}}, \frac{1}{\sqrt{2}}\right\} \\
r_n^L &= \left\{\frac{1}{\sqrt{2}}, \frac{1}{\sqrt{2}}\right\}, \qquad r_n^H = \left\{\frac{1}{\sqrt{2}}, -\frac{1}{\sqrt{2}}\right\}.
\end{aligned} \tag{2.110}
$$

The reader can verify that the Haar DWT can be obtained by the convolution of the filters above.

As the order increases, the length of the filters increases. The filters for the popular Daubechies-2(db_2) are as follows.

$$
\begin{aligned}
d_n^L &= \left\{\frac{1-\sqrt{3}}{4\sqrt{2}}, \frac{3-\sqrt{3}}{4\sqrt{2}}, \frac{3-\sqrt{3}}{4\sqrt{2}}, \frac{1+\sqrt{3}}{4\sqrt{2}}\right\} \\
d_n^H &= \left\{-\frac{1+\sqrt{3}}{4\sqrt{2}}, \frac{3-\sqrt{3}}{4\sqrt{2}}, -\frac{3-\sqrt{3}}{4\sqrt{2}}, \frac{1-\sqrt{3}}{4\sqrt{2}}\right\}
\end{aligned}
$$

$$r_n^L = \left\{\frac{1+\sqrt{3}}{4\sqrt{2}}, \frac{3-\sqrt{3}}{4\sqrt{2}}, \frac{3-\sqrt{3}}{4\sqrt{2}}, \frac{1-\sqrt{3}}{4\sqrt{2}}\right\}$$
$$r_n^H = \left\{\frac{1-\sqrt{3}}{4\sqrt{2}}, -\frac{3-\sqrt{3}}{4\sqrt{2}}, \frac{3-\sqrt{3}}{4\sqrt{2}}, -\frac{1+\sqrt{3}}{4\sqrt{2}}\right\}. \tag{2.111}$$

Because the wavelet basis functions are orthonormal, we can also think of DWT as an orthogonal transformation. Given a time series $\mathbf{x} = \mathbf{f}_j$ of length $n = 2^N$, its DWT coefficients are $\mathbf{d}_{j-1}, \mathbf{d}_{j-2}, \ldots, \mathbf{d}_J$ and $\mathbf{f}_J$. If we write these coefficients in vector form as

$$\mathbf{X} = (\mathbf{d}_{j-1}, \mathbf{d}_{j-2}, \ldots, \mathbf{d}_J, \mathbf{f}_J), \tag{2.112}$$

the DWT can be written as

$$\mathbf{X} = \mathbf{x}\mathbf{W}. \tag{2.113}$$

The size of $\mathbf{X}$ is also n. The transform matrix $\mathbf{W}$ is an $n \times n$ matrix. $\mathbf{W}$ is an orthonormal matrix. Each column of $\mathbf{W}$ is an orthonormal basis vector for $\mathbf{R}^n$.

Given the filters of a wavelet, we can compute its transform matrix $\mathbf{W}$. Considering a canonical basic vector of size n $\mathbf{e}_i = (0 \ \ldots \ 0 \ 1 \ 0 \ \ldots \ 0)$, where the i-th element is 1. Also we have

$$\mathbf{x} = \big(x(1), x(2), \ldots, x(n)\big) = \sum_{i=1}^{n} x(i)\mathbf{e}_i. \tag{2.114}$$

From the decomposition low-pass filter $\{d_n^L\}$ and decomposition high-pass filter $\{d_n^H\}$ of a Wavelet Transform, we can compute the DWT of $\mathbf{e}_i$, $c_i = (c_i(1), c_i(n), \ldots, c_i(n)) = DWT(\mathbf{e}_i)$. Obviously, the DWT is a linear transform, and therefore,

$$\mathbf{X} = DWT(\mathbf{x}) = \sum_{i=1}^{n} x(i) DWT(\mathbf{e}_i) = \sum_{i=1}^{n} x(i)\mathbf{c}_i. \tag{2.115}$$

In matrix form, this is

$$\mathbf{X} = (x(1), x(2), \ldots, x(n)) \begin{pmatrix} \mathbf{c}_1 \\ \mathbf{c}_2 \\ \vdots \\ \mathbf{c}_n \end{pmatrix} = \mathbf{x}\mathbf{W}, \tag{2.116}$$

where

$$\mathbf{W} = \begin{pmatrix} c_1(1) & c_1(2) & \ldots & c_1(n) \\ c_2(1) & c_2(2) & \ldots & c_2(n) \\ \vdots & \vdots & \ddots & \vdots \\ c_n(1) & c_n(2) & \ldots & c_n(n) \end{pmatrix}. \tag{2.117}$$

2.2.6 Wavelets – The Bottom Line

As we approximate a time series at higher and higher resolution, the wavelet coefficients of a time series in the higher resolution get closer and closer to zero. To use a Discrete Wavelet Transform as a data reduction tool for time series, we will keep only the DWT coefficients at the coarser approximation levels. Thus we will have a time series representation that can capture the trend of the time series.

In Figure 2.12 we show the approximation of a random walk time series with the Haar wavelet. The time series approximation at the top is reconstructed from the coarsest level of approximation and three levels of details. In the figure from top to bottom, the approximations have one more level of detail each.

In Figure 2.13 we show the approximation of a random walk time series with the db_2 wavelet. In contrast to the Haar wavelet, the approximation here is not piecewise constant. The approximation is smoother. In general, the db_2 wavelet can approximate continuous functions, and thus continuous time series, better than the Haar wavelet.

For the same wavelet family, as the order increase, the scaling and wavelet functions become more continuous. The approximation with a higher order wavelet is therefore more continuous. This is demonstrated in the following example. In Figure 2.14, we show the approximation of the same time series with another wavelet family: Coiflets wavelet.

As we can see from Figure 2.12, the approximation of a time series with the Haar wavelet results in a approximation that is a few segments of sequences of equal length, and each segment is approximated by the average value of the original time series that fall within the segments. This could be generalized to *Piecewise Aggregate Approximation (PAA)*, which has been proposed by Yi and Faloutsos [99], and Keogh et al. [55] independently.

The Piecewise Aggregate Approximation as a data reduction method will divide a time series of length n into k segments of equal length,[4] and keep the average values of each segment as the PAA coefficients. It is trivial to reconstruct the time series approximation giving the PAA coefficients. In Figure 2.15, we shows the basis vectors of a PAA that keep 8 segments for a time series of length 200.

The transformation based on PAA is very fast. PAA is easy to understand and implement. Simple as it is, PAA is very flexible in approximating the distance between time series.

Yi and Faloutsos [99] show that such a data reduction can be used for arbitrary L_p distance[5] approximation. Keogh et al. [55] show that PAA can be used for *weighted Euclidean distance*. Combining their results, we conclude that PAA can also used for *weighted L_p distance.*

---

[4] Padding zeroes at the end of the time series if necessary.

[5] L_p distance between time series $\mathbf{x}$ and $\mathbf{y}$ is defined as

$$||\mathbf{x} - \mathbf{y}||_p = (\sum_i |x_i - y_i|^p)^{1/p}.$$

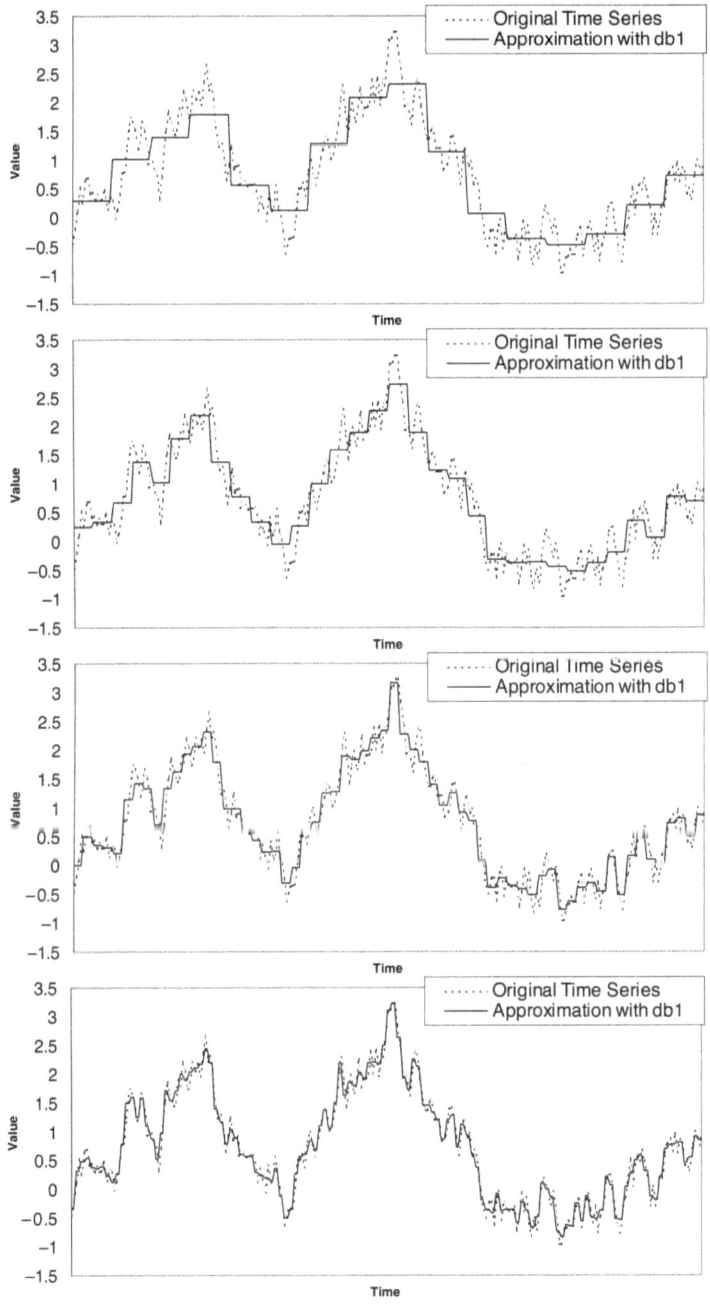

Figure 2.12. Approximations of a random walk time series with the Haar wavelet. From top to bottom are the time series approximations in resolution level 3, 4, 5 and 6 respectively.

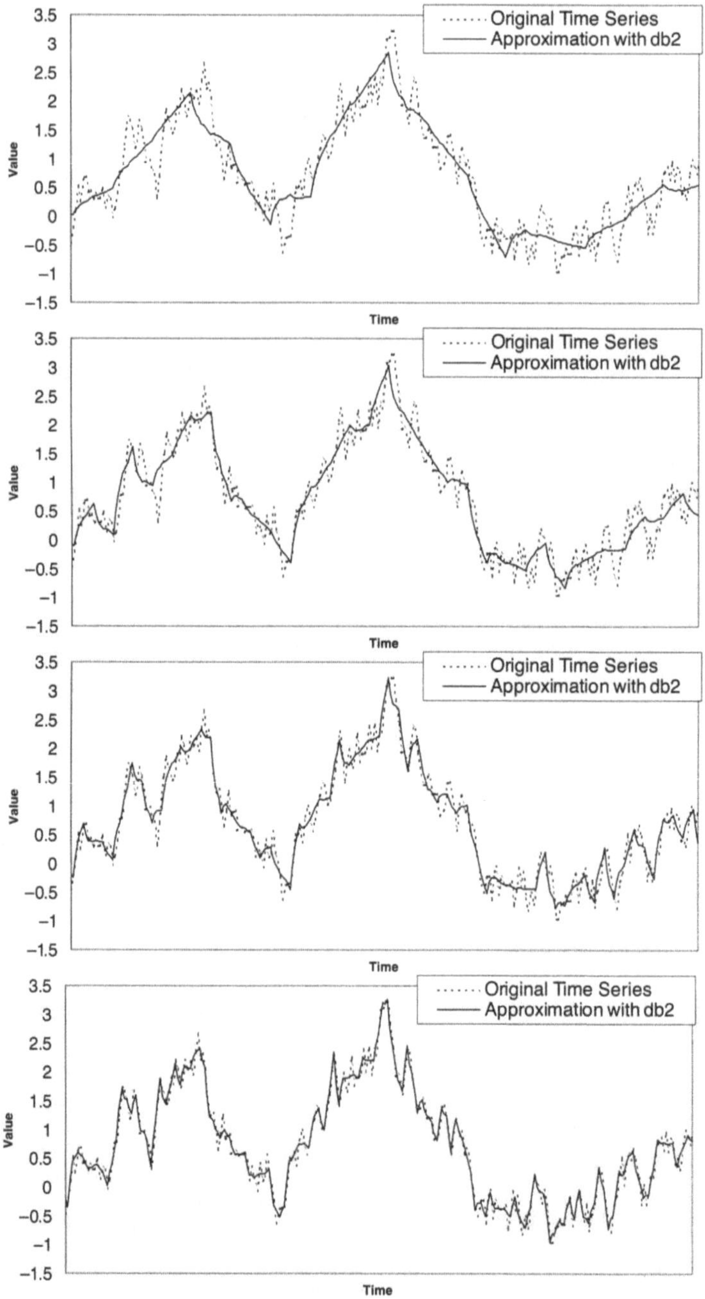

Figure 2.13. Approximations of a random walk time series with the db_2 wavelet. From top to bottom are the time series approximations in resolution level 3, 4, 5 and 6 respectively.

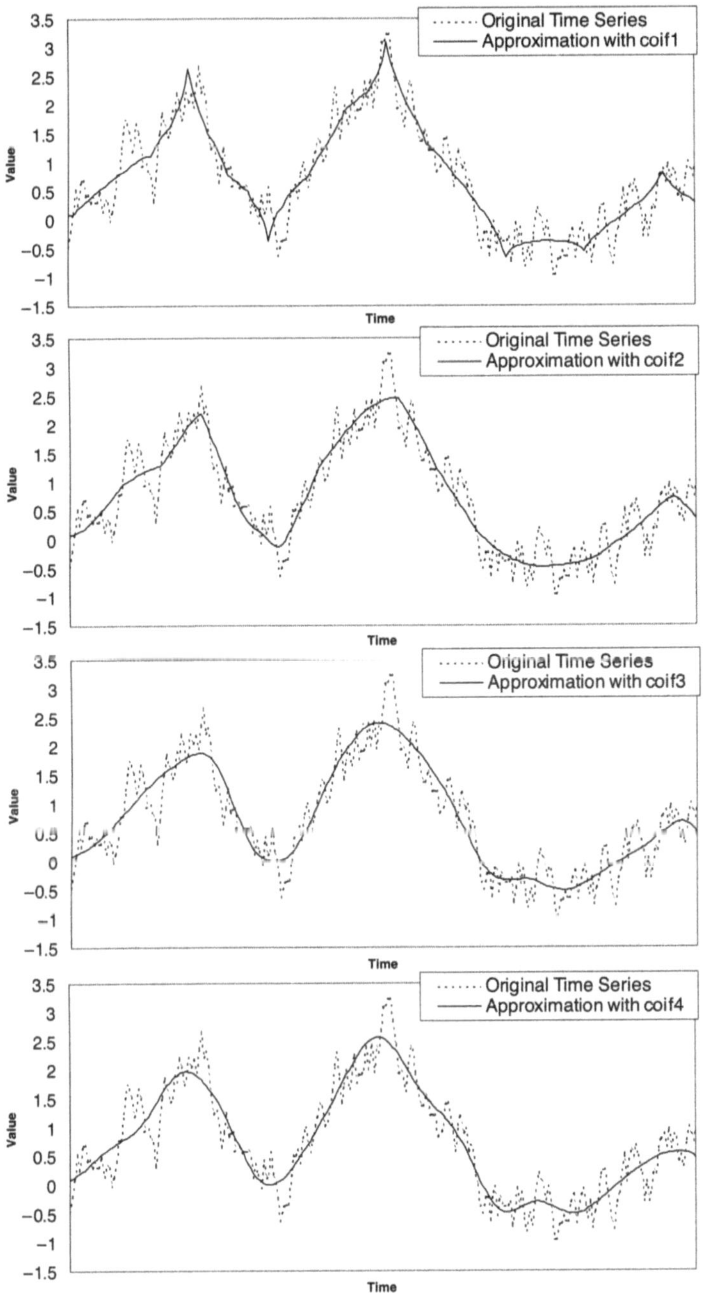

Figure 2.14. Approximations of a random walk time series with the coif wavelet family. From the top to the bottom, the time series is approximated by $coif_1$, $coif_2$, $coif_3$ and $coif_4$ respectively.

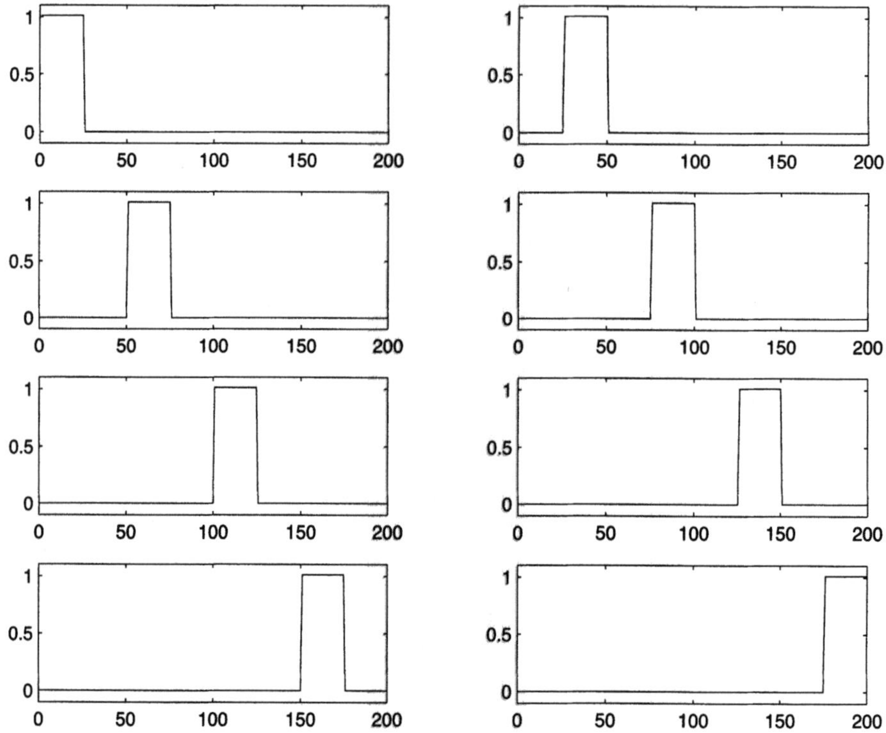

Figure 2.15. Basis vectors of a time series of size 200.

We have shown how to use different families of the Discrete Wavelet Transform to approximate a time series. In the above examples, the DWT coefficients we keep are the approximate wavelet coefficients in the highest level and the detail wavelet coefficients in the higher levels. Such an approximation catches the trends of the time series, but loses some detail information of the time series. Of course, this is not the only way to perform a DWT approximation.

Another way to perform a Discrete Wavelet Transform approximation is to keep the most significant DWT coefficients. In fact this is provably the optimal way to retain the energy of time series given a specific Wavelet Transform.

Remember that the Wavelet Transform is orthonormal, and it preserves the energy of the time series. Given a time series $\mathbf{x} = (x(1), x(2), \ldots, x(n))$ and its DWT coefficients $\mathbf{X} = (X(1), X(2), \ldots, X(n))$, we have

$$En(\mathbf{x}) = En(\mathbf{X}) = \sum_{i=1}^{n} X^2(n).$$

Therefore the best way to retain the energy of a time series with only k, $k < n$, DWT coefficients is to keep the k most significant DWT coefficients, i.e., the coefficients with the top k absolute values.

Figure 2.16 compares the Haar Wavelet approximations of an ECG time series with the first few coefficients and the most significant coefficients. We can see that by choosing the coefficients adaptively based on their significance, we can approximate the important details of the spikes in this example much better.

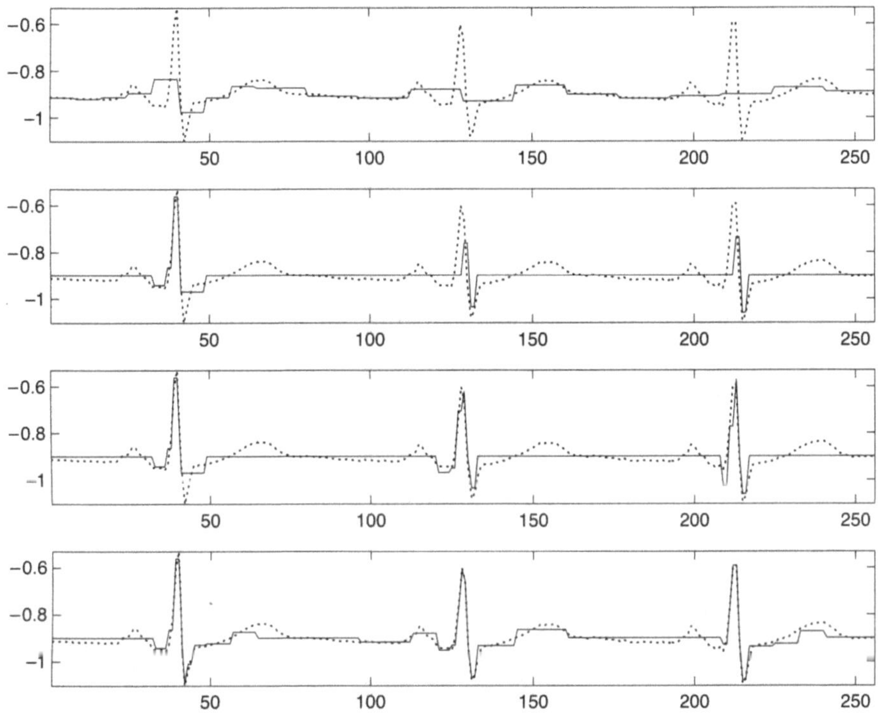

Figure 2.16. Approximation of an ECG time series with the DWT. From top to bottom, the time series is approximated by (a) the first 20 Haar coefficients; (b) the 5 most significant coefficients; (c) the 10 most significant coefficients; (d) the 20 most significant coefficients.

Of course the adaptive DWT approximation comes at a price. Because different time series have significant coefficients in different places, indexing time series with such dynamic DWT coefficients becomes much harder, though still possible [54].

As with the Fourier Transform, many operations are easier in Wavelet Space. For example, the Wavelet Transforms we have discussed in this section and the Fourier Transform are all orthonormal transforms. Such transforms preserve the L_2 distance between time series. And because there is a simple mapping relationship between the correlation and the L_2 distance as we will explain in later chapters, the correlation between time series is also preserved through Wavelet Transforms.

2.3 Singular Value Decomposition

The Discrete Fourier Transform (DFT) and the Discrete Wavelet Transform (DWT) of different wavelet families are all based on the orthogonal transform of time series. We choose the orthogonal basis vectors based on the nature of the time series and keep a few coefficients of the transform to approximate the original time series. One might wonder whether there is an optimal orthogonal transform that is the best in approximating the original time series with as few coefficients as possible. The answer is affirmative: the Singular Value Decomposition.

2.3.1 Existence and Uniqueness of Singular Value Decomposition

First we introduce the concept of *Singular Value Decomposition*. For more details about SVD, please refer to [79].

Consider a collection of m time series $\mathbf{x}_i, i = 1, 2, \ldots, m$, each time series has the same length n: $\mathbf{x}_i = (x_i(1), x_i(2), \ldots, x_i(n))$. We can write such a collection of time series in the following matrix form:

$$\mathbf{A}_{m \times n} = \begin{pmatrix} \mathbf{x}_1 \\ \mathbf{x}_2 \\ \vdots \\ \mathbf{x}_m \end{pmatrix} = \begin{pmatrix} x_1(1) & x_1(2) & \ldots & x_1(n) \\ x_2(1) & x_2(2) & \ldots & x_2(n) \\ \vdots & \vdots & \ddots & \vdots \\ x_m(1) & x_m(2) & \ldots & x_m(n) \end{pmatrix}. \tag{2.118}$$

If we define $\mathbf{S} \triangleq \mathbf{A}\mathbf{A}^T$, then $\mathbf{S}_{m \times m}$ is a *symmetric positive semidefinite* matrix. From the basic theorems in linear algebra, we know that for any symmetric matrix $\mathbf{S}$, there exist m linearly independent *eigenvectors* $\mathbf{U}_1, \mathbf{U}_2, \ldots, \mathbf{U}_m$ and m *eigenvalues* $\lambda_1, \lambda_2, \ldots, \lambda_m$, such that

$$\mathbf{S}\mathbf{U}_i = \lambda_i \mathbf{U}_i, \qquad i = 1, 2, \ldots, m. \tag{2.119}$$

The eigenvectors $\mathbf{U}_i, i = 1, 2, \ldots, m$ can be chosen to be orthonormal. Because $\mathbf{S}$ is positive semidefinite, the eigenvalues are non-negative: $\lambda_i \geq 0, i = 1, 2, \ldots, m$. We can also choose to make the eigenvalues non-decreasing:

$$\lambda_1 \geq \lambda_2 \geq \cdots \geq \lambda_m \geq 0.$$

Let the rank of $\mathbf{S}$ be d. Remember that the rank d of a matrix $\mathbf{S}$ is the number of linearly independent row vectors in $\mathbf{S}$. (Actually d is also the number of linearly independent column vectors in $\mathbf{S}$.) Obviously, we have $d \leq m$. If $\mathbf{S}$ is singular, i.e., $d < m$, then

$$\lambda_{d+1} = \lambda_{d+2} = \cdots = \lambda_m = 0.$$

We also define a collection of vectors $\{\mathbf{V}_i\}$ as follows:

$$\mathbf{V}_i = \frac{1}{\sqrt{\lambda_i}} \mathbf{A}^T \mathbf{U}_i, \qquad i = 1, 2, \ldots, d. \tag{2.120}$$

The following properties of $\{\mathbf{V}_i\}$ and $\{\mathbf{U}_i\}$ are not hard to prove using elementary linear algebra.

Lemma 2.39. $\{\mathbf{V}_i\}_{i=1,2,\ldots,d}$ *are orthonormal.*

Proof.

$$\begin{aligned}\langle \mathbf{A}^T\mathbf{U}_i, \mathbf{A}^T\mathbf{U}_j\rangle &= \mathbf{U}_i^T\mathbf{A}\mathbf{A}^T\mathbf{U}_j = \mathbf{U}_i^T\mathbf{S}\mathbf{U}_j \\ &= \lambda_j\mathbf{U}_i^T\mathbf{U}_j = \lambda_j\langle \mathbf{U}_i, \mathbf{U}_j\rangle.\end{aligned}$$

Because $\{\mathbf{U}_i\}_{i=1,2,\ldots,d}$ are orthonormal, $\{\mathbf{V}_i\}_{i=1,2,\ldots,d}$ are also orthonormal. □

Lemma 2.40. $\mathbf{A}^T\mathbf{U}_i = \mathbf{0}_n$ *for* $i > d$*, where* $\mathbf{0}_n$ *is zero vector of size n.*

Proof. Because $i > d$, we have $\lambda_i = 0$.

$$||\mathbf{A}^T\mathbf{U}_i||^2 = \langle \mathbf{A}^T\mathbf{U}_i, \mathbf{A}^T\mathbf{U}_i\rangle = \lambda_i\langle \mathbf{U}_i, \mathbf{U}_i\rangle = 0.$$

Therefore,

$$\mathbf{A}^T\mathbf{U}_i = \mathbf{0}_n.$$ □

Lemma 2.41. $\mathbf{V}_i$ *is the eigenvector of the matrix* $\mathbf{A}^T\mathbf{A}$ *with the eigenvalue* λ_i*, i.e.,*

$$\mathbf{A}^T\mathbf{A}\mathbf{V}_i = \lambda_i\mathbf{V}_i, \qquad i = 1, 2, \ldots, d. \tag{2.121}$$

Proof.

$$\begin{aligned}\mathbf{A}^T\mathbf{A}\mathbf{V}_i &= \frac{1}{\sqrt{\lambda_i}}\mathbf{A}^T\mathbf{A}\mathbf{A}^T\mathbf{U}_i = \frac{1}{\sqrt{\lambda_i}}\mathbf{A}^T\mathbf{S}\mathbf{U}_i \\ &= \sqrt{\lambda_i}\mathbf{A}^T\mathbf{U}_i = \lambda_i\mathbf{V}_i.\end{aligned}$$ □

Lemma 2.42.

$$\mathbf{U}_i = \frac{1}{\sqrt{\lambda_i}}\mathbf{A}\mathbf{V}_i, \qquad i = 1, 2, \ldots, d. \tag{2.122}$$

Proof.

$$\mathbf{A}\mathbf{V}_i = \frac{1}{\sqrt{\lambda_i}}\mathbf{A}\mathbf{A}^T\mathbf{U}_i = \frac{1}{\sqrt{\lambda_i}}\mathbf{S}\mathbf{U}_i = \sqrt{\lambda_i}\mathbf{U}_i.$$ □

With the above lemmas, we are now ready to prove the following theorem that states the existence of Singular Value Decomposition for any matrix.

Theorem 2.43 (Existence of Singular Value Decomposition).

$$\mathbf{A} = \sum_{i=1}^{d}\sqrt{\lambda_i}\mathbf{U}_i\mathbf{V}_i^T \tag{2.123}$$

Proof. Because the eigenvectors $\{\mathbf{U}_i\}_{i=1,2,\ldots,m}$ are orthonormal, we have

$$\mathbf{I}_{m\times m} = \sum_{i=1}^{m} \mathbf{U}_i\mathbf{U}_i^T. \tag{2.124}$$

Therefore,

$$\begin{aligned}
\mathbf{A} &= \mathbf{I}_{m\times m}\mathbf{A} \\
&= \sum_{i=1}^{m} \mathbf{U}_i\mathbf{U}_i^T\mathbf{A} \\
&= \sum_{i=1}^{m} \mathbf{U}_i(\mathbf{A}^T\mathbf{U}_i)^T \\
&= \sum_{i=1}^{d} \sqrt{\lambda_i}\mathbf{U}_i\mathbf{V}_i^T + \sum_{i=d+1}^{m} \mathbf{U}_i(\mathbf{A}^T\mathbf{U}_i)^T \quad \text{(Lemma 2.40)} \\
&= \sum_{i=1}^{d} \sqrt{\lambda_i}\mathbf{U}_i\mathbf{V}_i^T.
\end{aligned}$$

□

Equation (2.123) is called the ***singular value decomposition (SVD)*** of the matrix **A**. $\sqrt{\lambda_i}, i = 1, 2, \ldots, d$ are the ***singular values*** of **A**. The vectors $\{\mathbf{U}_i\}_{i=1,2,\ldots,d}$ and $\{\mathbf{V}_i\}_{i=1,2,\ldots,d}$ are the *left singular vectors* and *right singular vectors* of the decomposition respectively.

The singular value decomposition of a matrix **A** always exists, no matter how singular **A** is.

The singular value decomposition of a matrix **A** is "almost" unique. Of course, if $\lambda_i = \lambda_{i+1}$, $(\mathbf{V}_i, \mathbf{U}_i)$ and $(\mathbf{V}_{i+1}, \mathbf{U}_{i+1})$ can be interchanged in (2.123). Also, if $(\mathbf{V}_i, \mathbf{U}_i)$ is a singular vector pair, then it can be replaced by $(-\mathbf{V}_i, -\mathbf{U}_i)$ in (2.123) too. The SVD of a matrix **A** is unique in the following sense:

- The singular values are uniquely decided by **A**.
- If all the singular values are different, the corresponding singular vector pairs are uniquely decided if we choose the sign of one vector of the singular vector pair.

You can find the proof of the uniqueness of SVD in the clearly written reference [40].

The Singular Value Decomposition (2.123) can also be written in the following matrix form.

$$\mathbf{A} = \mathbf{U}\Lambda\mathbf{V}^T, \tag{2.125}$$

where

$$\mathbf{U} = (\mathbf{U}_1, \mathbf{U}_2, \ldots, \mathbf{U}_d),\ \Lambda = \begin{pmatrix} \sqrt{\lambda_1} & & & \\ & \sqrt{\lambda_1} & & \\ & & \ddots & \\ & & & \sqrt{\lambda_d} \end{pmatrix},\ \mathbf{V}^T = \begin{pmatrix} \mathbf{V}_1^T \\ \mathbf{V}_2^T \\ \vdots \\ \mathbf{V}_d^T \end{pmatrix}. \tag{2.126}$$

Remember that **A** is a $m \times n$ matrix, $\mathbf{U}_i$ is column vector of size m and $\mathbf{V}_i$ is column vector of size n. Therefore **U** is a $m \times d$ matrix, Λ is a $d \times d$ matrix, and $\mathbf{V}^T$ is a $d \times n$ matrix. The following equation shows the matrix multiplication visually.

$$\left(\quad \mathbf{A}_{m\times n} \quad \right)$$

$$= \left(\quad \mathbf{U}_{m\times d} \quad \right) \left(\quad \Lambda_{d\times d} \quad \right) \left(\quad \mathbf{V}^T_{d\times n} \quad \right).$$

As a toy example of Singular Value Decomposition, consider the four time series (1, 1, 1, 0, 0), (2, 2, 2, 0, 0), (0, 0, 0, 1, 1) and (0, 0, 0, 2, 2). The matrix form of the collection of time series is

$$\mathbf{A} = \begin{pmatrix} 1\,1\,1\,0\,0 \\ 2\,2\,2\,0\,0 \\ 0\,0\,0\,1\,1 \\ 0\,0\,0\,2\,2 \end{pmatrix}.$$

To compute the Singular Value Decomposition of **A**, we have

$$\mathbf{S} = \mathbf{A}\mathbf{A}^T = \begin{pmatrix} 3 & 6 & 0\,0 \\ 6 & 12 & 0\,0 \\ 0 & 0 & 2\,4 \\ 0 & 0 & 4\,8 \end{pmatrix}.$$

The eigenvalues of **S** are 15, 10, 0 and 0. The normalized eigenvector for $\lambda_1 = 15$ is $\mathbf{U}_1 = \frac{1}{\sqrt{5}}(1, 2, 0, 0)^T$. The normalized eigenvector for $\lambda_2 = 10$ is $\mathbf{U}_2 = \frac{1}{\sqrt{5}}(0, 0, 1, 2)^T$. From (2.120) we have

$$\mathbf{V}_1^T = \frac{1}{\sqrt{3}}(1, 1, 1, 0, 0),$$

and

$$\mathbf{V}_2^T = \frac{1}{\sqrt{2}}(0, 0, 0, 1, 1).$$

So the SVD of **A** is

$$\mathbf{A} = \begin{pmatrix} \frac{1}{\sqrt{5}} & 0 \\ \frac{2}{\sqrt{5}} & 0 \\ 0 & \frac{1}{\sqrt{5}} \\ 0 & \frac{2}{\sqrt{5}} \end{pmatrix} \begin{pmatrix} \sqrt{15} & 0 \\ 0 & \sqrt{10} \end{pmatrix} \begin{pmatrix} \frac{1}{\sqrt{3}} & \frac{1}{\sqrt{3}} & \frac{1}{\sqrt{3}} & 0 & 0 \\ 0 & 0 & 0 & \frac{1}{\sqrt{2}} & \frac{1}{\sqrt{2}} \end{pmatrix}.$$

2.3.2 Optimality of Singular Value Decomposition

The Singular Value Decomposition is optimal in minimizing the *matrix norm*. Here we give the definition of matrix norm, which is also known as *Frobenius matrix norm*.

Definition 2.44 (Matrix norm). *The matrix norm of a matrix* $\mathbf{A} = \{a_{ij}\}_{i,j=1}^{m,n}$ *is defined as*

$$||\mathbf{A}|| = \sqrt{\sum_{i=1}^{m}\sum_{j=1}^{n} a_{ij}^2}. \tag{2.127}$$

Obviously, the total energy of the a collection of time series $\mathbf{x}_i$, $i = 1, 2, \ldots, m$, is the square of the norm of its corresponding matrix in (2.118):

$$\sum_{i=1}^{m} En(\mathbf{x}_i) = ||\mathbf{A}||^2. \tag{2.128}$$

The following theorem states the optimal feature of SVD. For its proof, please consult [40].

Theorem 2.45 (Optimality of Singular Value Decomposition). *Let* $\mathbf{P}_i$ *be an arbitrary column vector of size* m *and* $\mathbf{Q}_i$ *an arbitrary column vector of size* n*; given a matrix* $\mathbf{A}_{m\times n}$ *with rank* d *and a integer* $k < d$*, then for all the matrices of the form*

$$\mathbf{B} = \sum_{i=1}^{k} \mathbf{P}_i \mathbf{Q}_i^T, \tag{2.129}$$

we have

$$\min_{\mathbf{B}} ||\mathbf{A} - \mathbf{B}||^2 = \sum_{i=k+1}^{d} \lambda_i, \tag{2.130}$$

where the minimum is achieved with $\mathbf{B}_0$*:*

$$||\mathbf{A} - \mathbf{B_0}||^2 = \min_{\mathbf{B}} ||\mathbf{A} - \mathbf{B}||^2,$$

$$\mathbf{B_0} = \sum_{i=1}^{k} \sqrt{\lambda_i} \mathbf{U}_i \mathbf{V}_i^T. \tag{2.131}$$

2.3.3 Data Reduction in Singular Value Decomposition

Suppose that we have a collection of time series $\mathbf{x}_i, i = 1, 2, \ldots, m$ and k orthonormal basis vectors $\mathbf{Q}_j, j = 1, 2, \ldots, k$, where the time series and the basis vectors have the length n. We can represent the time series as

$$\mathbf{x}_i = \sum_{j=1}^{k} c_{ij} \mathbf{Q}_j^T + R(\mathbf{x}_i), \qquad i = 1, 2, \ldots, m, \tag{2.132}$$

where $R(\mathbf{x}_i)$ is the residual vector of $\mathbf{x}_i$ when it is decomposed in the space spanned by $\{\mathbf{Q}_j\}_{j=1,2,\ldots,k}$. Obviously, $R(\mathbf{x}_i)$ is orthogonal to $\{\mathbf{Q}_j\}_{j=1,2,\ldots,k}$.

On both sides of (2.132), if we compute the inner product with $\mathbf{Q}_j$, $j = 1, 2, \ldots, k$, we have

$$c_{ij} = \langle \mathbf{x}_i, \mathbf{Q}_j^T \rangle, \qquad i = 1, 2, \ldots, m, \ j = 1, 2, \ldots, k. \tag{2.133}$$

Let $\tilde{\mathbf{x}}_i = \mathbf{x}_i - R(\mathbf{x}_i)$ be the approximation of $\mathbf{x}_i$; we have

$$\mathbf{x}_i \approx \tilde{\mathbf{x}}_i = \mathbf{x}_i - R(\mathbf{x}_i) = \sum_{j=1}^{k} c_{ij} \mathbf{Q}_j^T, \qquad i = 1, 2, \ldots, m. \tag{2.134}$$

What is the best selection of the orthonormal basis vectors, $\{\mathbf{Q}_j\}_{j=1,2,\ldots,k}$, such that the energy of the collection of time series is preserved as much as possible, i.e., the total approximation errors are as small as possible? Let $\mathbf{P}_j = (c_{1j}, \ldots, c_{mj})^T$, $j = 1, 2, \ldots, k$; the total approximation errors are

$$\sum_{i=1}^{m} ||\mathbf{x}_i - \tilde{\mathbf{x}}_i||^2 = \sum_{i=1}^{m} ||\mathbf{x}_i - \sum_{j=1}^{k} c_{ij} \mathbf{Q}_j^T ||^2$$
$$= ||\mathbf{A} - \sum_{j=1}^{k} \mathbf{P}_j \mathbf{Q}_j^T ||^2.$$

From Theorem 2.45, if we choose

$$\mathbf{Q}_i = \mathbf{V}_i, \qquad i = 1, 2, \ldots, k,$$

we can obtain an optimal approximation that preserves the energy of a collection of time series.

The above observation yields the SVD data reduction for time series. For a collection of time series, $\mathbf{x}_i, i = 1, 2, \ldots, m$, and its matrix form $\mathbf{A}$, the rank of $\mathbf{A}$ is usually close to $\min(m, n)$. However, λ_i decreases as i increases. It is very likely that the energy of a collection of time series is concentrated in the first few eigenvalues $\sum_{i=1}^{k} \lambda_i$, $k \ll n$. We can thus use only the first k SVD coefficients to approximate a time series. That is,

$$\mathbf{A} \approx \hat{\mathbf{U}}\hat{\Lambda}\hat{\mathbf{V}}^T, \tag{2.135}$$

where

$$\hat{\mathbf{U}} = (\mathbf{U}_1, \mathbf{U}_2, \ldots, \mathbf{U}_k),\ \hat{\Lambda} = \begin{pmatrix} \sqrt{\lambda_1} & & & \\ & \sqrt{\lambda_1} & & \\ & & \ddots & \\ & & & \sqrt{\lambda_k} \end{pmatrix},\ \hat{\mathbf{V}}^T = \begin{pmatrix} \mathbf{V}_1^T \\ \mathbf{V}_2^T \\ \vdots \\ \mathbf{V}_k^T \end{pmatrix}. \tag{2.136}$$

2.3.4 Singular Value Decomposition – The Bottom Line

Here are the steps for using SVD for a collection of time series. First we write the collection of time series in the matrix form $\mathbf{A}$ (2.118) and perform the Singular Value Decomposition for $\mathbf{A}$. Most mathematical software, e.g., LAPACK [4], Matlab [3], have implemented SVD.

Using $\{\mathbf{V}_i\}_{i=1,2,\ldots,k}$ as the basis vectors, the SVD coefficients for all the time series in $\mathbf{A}$ is

$$\mathbf{C} = \mathbf{A}\hat{\mathbf{V}}, \tag{2.137}$$

where the SVD coefficients of time series $\mathbf{x}_j$ is the j-th row of $\mathbf{C}$.

Given any time series $\mathbf{y}$ of length n, if we want to compute its SVD coefficient $\mathbf{Y}$ based on $\mathbf{A}$, i.e., to express $\mathbf{y}$ as the linear combination of $\{\mathbf{V}_i\}_{i=1,2,\ldots,k}$, from (2.133) we can compute it as

$$\mathbf{Y} = \mathbf{y}\hat{\mathbf{V}}. \tag{2.138}$$

Similarly, the reconstruction of $\tilde{\mathbf{y}}$, the approximation of a time series $\mathbf{y}$, from its SVD coefficients $\mathbf{Y}$ is given as in (2.134),

$$\tilde{\mathbf{y}} = \mathbf{Y}\hat{\mathbf{V}}^T. \tag{2.139}$$

The orthonormal basis vectors $\{\mathbf{V}_i\}$ of SVD allow us to minimize the approximation errors for a collection of time series globally. In DFT and DWT, the orthonormal basis vectors are data independent, that is, the basis vectors chosen are not derived from the data. By contrast, SVD finds the optimal basis vectors given the time series.

Here is an experiment on a collection of 300 random walk time series that demonstrates the above properties of SVD. Figure 2.17(a) shows two of the random walk time series and their SVD approximations. Each of the time series has the length

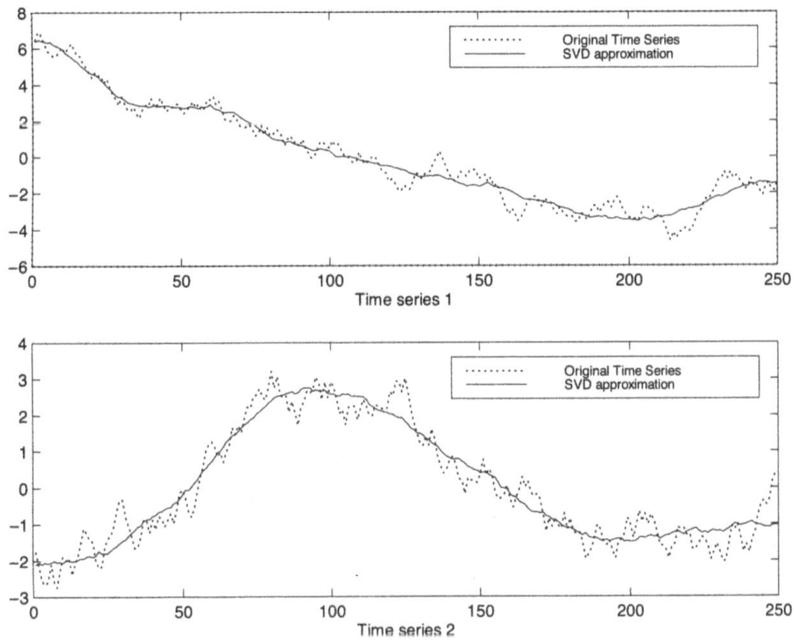

(a) Time Series 1 and 2 and their SVD approximations.

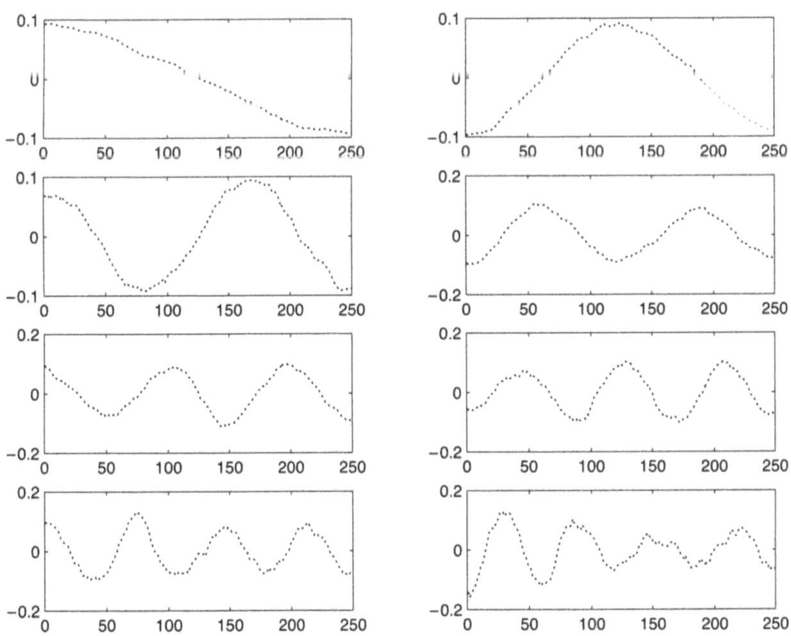

(b) The SVD basis vectors

Figure 2.17. SVD for a collection of random walk time series.

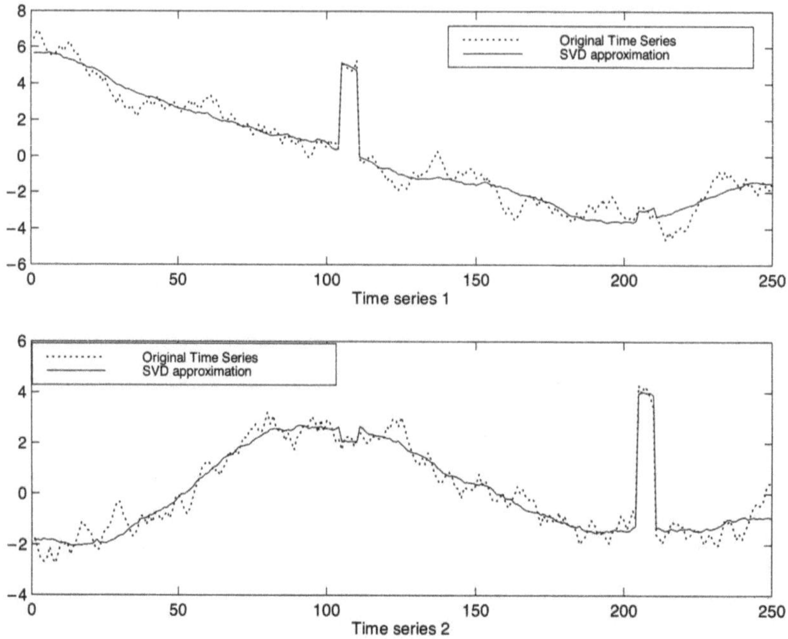

(a) Time Series 1 and 2 and their SVD approximations.

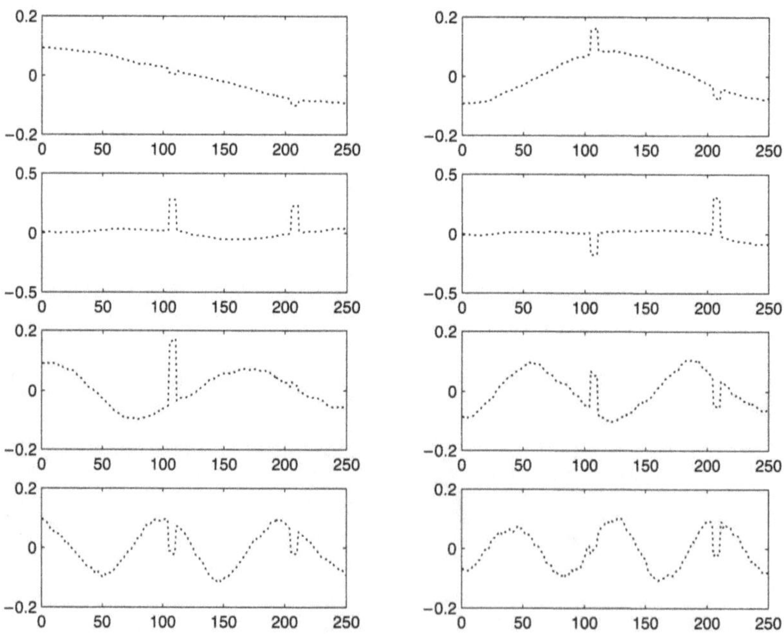

(b) The SVD basis vectors.

Figure 2.18. SVD for a collection of random walk time series with bursts.

$n = 250$. They are approximated by 8 SVD coefficients. We can see that with only 8 SVD coefficients, we capture the raw shape of the time series very well. The basis vectors for the SVD are also shown in Figure 2.17(b). We can see that the basis vectors are similar to the basis vectors of DFT, though they are not exactly trigonometric functions.

To demonstrate the adaptivity of SVD, we add some components to each of the time series in the collection. A short burst is superimposed to each time series either around the time point of 100 or 200. Two of the resulting time series samples are shown in Figure 2.18(a). We compute the SVD of the new collection of time series again and show their SVD approximations. The SVD approximations follow the new burst very closely. The reason is that the basis vectors computed from the new data can incorporate the burst adaptively. In Figure 2.18(b), the basis SVD vectors for the new collection of time series show the new burst components clearly.

SVD approximations are closer approximations to the real time series in terms of Euclidean distance than are Discrete Fourier Transforms or Discrete Wavelet Transforms. This means that using the same number of coefficients to approximate the correlation between time series, on average, we will expect the SVD to give the best approximation. In cases where the basis vectors found by the Discrete Fourier Transform (or the Discrete Wavelet Transform) are optimal, Singular Value Decomposition will find them too as Theorem 2.45 states and Figure 2.17(b) demonstrates.

2.4 Sketches

The data reduction techniques we have discussed so far are all based on orthogonal transformations. If we think of a time series as a point in some high dimensional coordinate space, an orthogonal transformation is just a rotation of the axes of the coordinate space. If a collection of time series have some principal components and the transformed axes are along these principal components, we can keep only those axes corresponding to the principal components to reduce the dimensionality of the time series data.

But what if the data do not have any clear principal components? Consider a collection of time series of white noise for example. Clearly, we cannot use orthogonal transformations such as DFT, DWT or SVD. This section introduces a new kind of Data Reduction technique: *random projection*.

In random projection, we try to approximate the distance pairs of time series by comparing each member of that pair to a reference collection of random time series (called a sketch pool). If two time series have similar distances to the corresponding members of the sketch pool, then they are likely to be similar to one another.

We'll explain why this works through an analogy. Suppose that you are lost in the forest and you have only an old-fashioned cell phone without a Global Positioning System (GPS). Luckily, you are in the covered area of your mobile service provider and you can call a friend who may be near by. By knowing how far you are away from some landmarks, say 2 miles from a hill, 50 yards from a creek, etc., and getting the same information from your friend, you may be able to determine that the two of you are close to one another. Your position in the forest is the time series in the high dimensional space, the random vectors in the sketch pool serve as the random landmarks and the sketches are just your distances from these landmarks.

Unlike data reduction based on orthogonal transformations, random projection can approximate different types of distances. We will start with the Euclidean Distance.

2.4.1 Euclidean Distance

Let us start by giving a formal definition of sketch pools.

Definition 2.46 (Sketches). *Given a time series* $\mathbf{x}^n = (x(1), x(2), \ldots, x(n))$ *and a collection of* k *vectors* $\mathbf{r}_i^n = (r_i(1), r_i(2), \ldots, r_i(n)), i = 1, 2, \ldots, k$, *where all elements* $r_i(j), i = 1, 2, \ldots, k, j = 1, 2, \ldots, n$, *are random variables from a distribution* $\mathcal{D}$, *the* $\mathcal{D}$*-sketches of* $\mathbf{x}$ *are defined as* $\mathbf{s}(\mathbf{x}) = (s(1), s(2), \ldots, s(k))$, *where*

$$s(i) = \langle \mathbf{x}, \mathbf{r}_i \rangle, \qquad i = 1, 2, \ldots, k, \tag{2.140}$$

i.e., $s(i)$ *is the inner product between* $\mathbf{x}$ *and* $\mathbf{r}_i$.

The sketches can be written as

$$\mathbf{s}(\mathbf{x}) = \mathbf{x}\mathbf{R}, \tag{2.141}$$

where

$$\mathbf{R}_{n \times k} = (\mathbf{r}_1^T, \mathbf{r}_2^T, \ldots, \mathbf{r}_k^T) = \begin{pmatrix} r_1(1) & r_2(1) & \cdots & r_k(1) \\ r_1(2) & r_2(2) & \cdots & r_k(2) \\ \vdots & \vdots & \ddots & \vdots \\ r_1(n) & r_2(n) & \cdots & r_k(n) \end{pmatrix}. \tag{2.142}$$

The collection of random vectors $\mathbf{R}$ is called the *sketch pool*. Obviously, the time complexity to compute the sketch for each time series is $O(nk)$.

If the distribution $\mathcal{D}$ is a Gaussian distribution, we can compute the Gaussian sketches of a time series. The most important property of sketches is stated by the Johnson–Lindenstrauss lemma [52] for Gaussian sketches.

Lemma 2.47. *Given a collection* C *of* m *time series with length* n, *for any two time series* $\mathbf{x}, \mathbf{y} \in C$, *if* $\epsilon < 1/2$ *and* $k = \frac{9 \log m}{\epsilon^2}$, *then*

$$(1 - \epsilon) \leq \frac{||\mathbf{s}(\mathbf{x}) - \mathbf{s}(\mathbf{y})||^2}{||\mathbf{x} - \mathbf{y}||^2} \leq (1 + \epsilon) \tag{2.143}$$

holds with probability $1/2$, *where* $\mathbf{s}(\mathbf{x})$ *is the Gaussian sketches of* $\mathbf{x}$.

In other words, the Johnson–Lindenstrauss lemma says that a collection of m points in a n-dimensional space can be mapped to a k-dimensional sketch space. The Euclidean distances between any pair of points in the sketch space approximate their true distance in the n-dimensional space with probability $1/2$. There are also other flavors of random projection based on the Johnson–Lindenstrauss lemma that give a similar probabilistic approximation of Euclidean distance, such as [7].

The lemma may sound like magic. How can we have probabilistic approximation of the Euclidean distance between time series from the sketches, which are computed from random vectors? The intuition is based on the forest analogy above. If you and your friend are both two kilometers from the tall tree, you may be four kilometers apart, but if you are also both two kilometers from the big rock and the rock is far from the tree, then you are likely to be close to one another.

From the lemma, we can see that if we increase the number of sketches k, the approximation error ϵ will be smaller. Also we can boost the probability of success using standard randomization methods. If a sketch approximation is within ϵ, we call it a success. We keep w different sketches and repeat the approximate distance computation w times; the probability that the median of the approximate distances is within precision ϵ is the same as the probability that the number of success is larger than $w/2$. From the Chernoff bound, if we compute $O(\log 1/\delta)$ repetition of sketches and take the median of the distance between the sketches, the probability of success can be boosted to $1 - \delta$.

When using sketches to approximate the Euclidean distances between time series, we do not require the time series to have principal components. We perform the following experiments to verify this.

The time series under consideration is a collection of $10,000$ stock price time series. The time unit of the time series is one second. Each time series in the collection has a size $3,600$, corresponding to one hour's data. We compute the sketches of these time series with sketch pools of size $k = 40$ and therefore reduce the dimensionality of the time series from $3,600$ to 40. We randomly pick $1,000$ pairs of time series in the collection and compute their Euclidean distances. The approximations of the distances using sketches are also computed. Let the approximation error be the ratio of the absolute difference between the approximate distance and the true distance to the true distance. The distributions of approximation errors using sketches are shown in Figure 2.19(a). For example, from the figure we can see that 90% of the approximation errors are within 0.32. We also repeat the experiment for larger sketch sizes $k = 80$ and $k = 160$. Larger sketch sizes give better approximations. For example, with $k = 160$, 90% of the approximation errors are within 0.22, while with $k = 80$, 90% of the approximation errors are within 0.19.

Whereas stock price data modeled by a random walk has a few large principal components, the price return time series is best modeled as white noise. The price return time series is defined as the time series derived from the price time series by computing the point-by-point price differences, perhaps divided by the price. That is, given a price time series $\mathbf{x}^n = (x(1), x(2), \ldots, x(n))$, its price return time series is $\mathbf{dx}^n = (x(2) - x(1), x(3) - x(2), \ldots, x(n) - x(n-1))$. There is no principal component in the price return time series. Will the distance approximations using

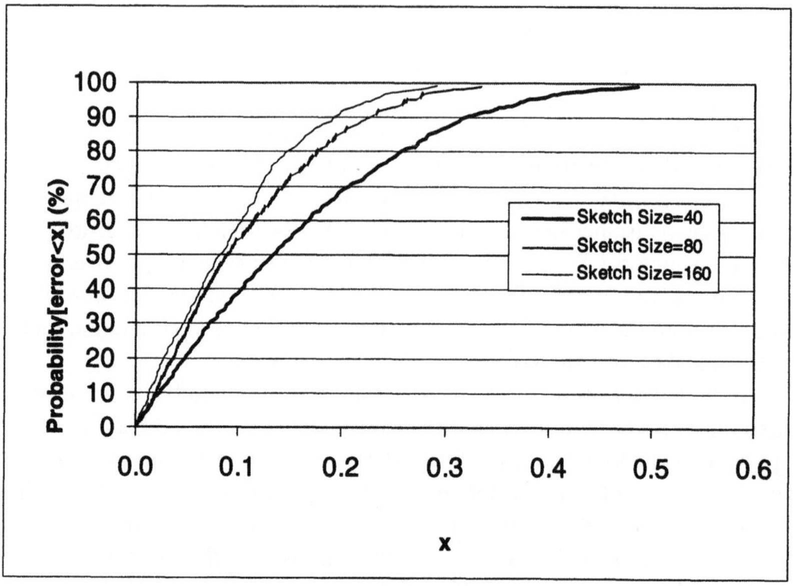

(a) Price time series

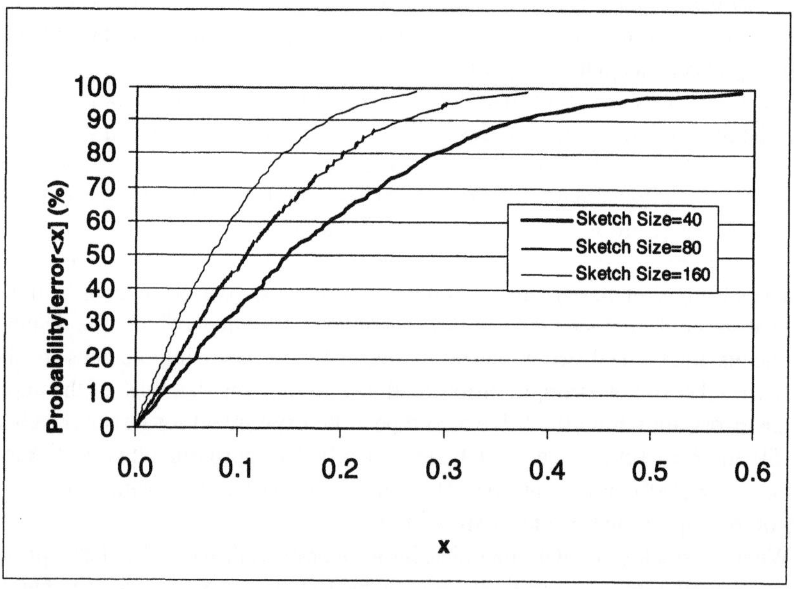

(b) Price return time series

Figure 2.19. The approximation of distances between time series using sketches; 1 hour of stock data.

sketches work for these price return time series? We repeat the previous experiment on the price return time series. The results are shown in Figure 2.19(b). We can see that the qualities of approximations using sketches are very close for both data sets.

One interesting observation is that the sketch size k depends only on the number of time series m, the approximation bound ϵ and the probability guarantee bound δ. The sketch size k does not depend on the length of the time series n. This makes random projections ideal for data reduction for a relatively small collection of time series with very long length. We repeat the previous experiment on a collection of stock price time series with longer lengths. The size of the collection of time series is the same as before, but the length of each time series in the collection is doubled, corresponding to two hours' data. From Figure 2.20, we can see that the same quality of approximation using sketches is achieved even though the time series are longer.

The number of sketches may have to be large if the approximation requirement is demanding (small ϵ and δ). We can use the SVD to further reduce the dimensions of the sketch space. This works especially well if the time series data have some principal components after the random projection.

2.4.2 L_p Distance

In addition to approximating Euclidean distances, sketches can be used to approximate L_p-distance. Although Euclidean distance is used most often for time series, other distance measures between time series can provide interesting results too.

The approximation of L_p-distance is based on the concept of *stable distribution* [76]. A stable distribution $\mathcal{D}(p)$ is a statistical distribution with parameter $p \in (0, 2]$ having the property:

Definition 2.48 (Stable distribution). *A distribution $\mathcal{D}(p)$ is stable if for any n real number $a_1, a_2, \ldots, a_n$ and n i.i.d. (independent and identically distributed) random variables $X_1, X_2, \ldots, X_n$ from $\mathcal{D}(p)$,*

$$\sum_{i=1}^{n} a_i X_i \sim \Big(\sum_{i=1}^{n} |a_i|^p\Big)^{1/p} X, \tag{2.144}$$

i.e., $\sum_{i=1}^{n} a_i X_i$ has the same distribution as $(\sum_{i=1}^{n} |a_i|^p)^{1/p} X$, where X is drawn from $\mathcal{D}(p)$.

$\mathcal{D}(2)$ is a Gaussian distribution and $\mathcal{D}(1)$ is Cauchy distribution. Indyk [46] shows that one can construct $\mathcal{D}(p)$ sketches to approximate L_p-distance.

Lemma 2.49. *Given a collection C of m time series with length n, for any two time series $\mathbf{x}, \mathbf{y} \in C$, if $k = c\frac{\log(1/\delta)}{\epsilon^2}$ for some constant c, let $\mathbf{s}(\mathbf{x})$ and $\mathbf{s}(\mathbf{y})$be the $\mathcal{D}(p)$ sketches of $\mathbf{x}$ and $\mathbf{y}$ with size k; then*

$$(1-\epsilon) \le \frac{B(p)\,\text{median}\,(|\mathbf{s}(\mathbf{x}) - \mathbf{s}(\mathbf{y})|)}{||\mathbf{x}-\mathbf{y}||_p} \le (1+\epsilon) \tag{2.145}$$

holds with probability $1-\delta$, where $B(p)$ is some scaling factor.

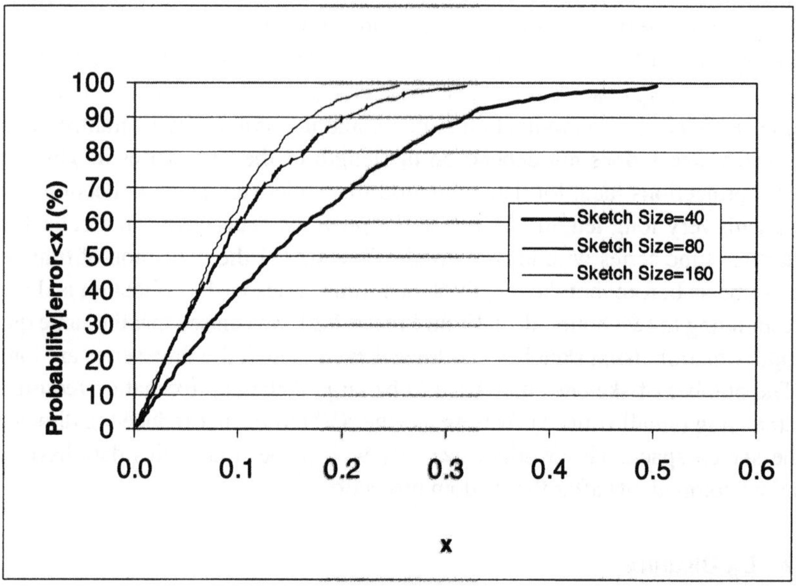

(a) Price time series

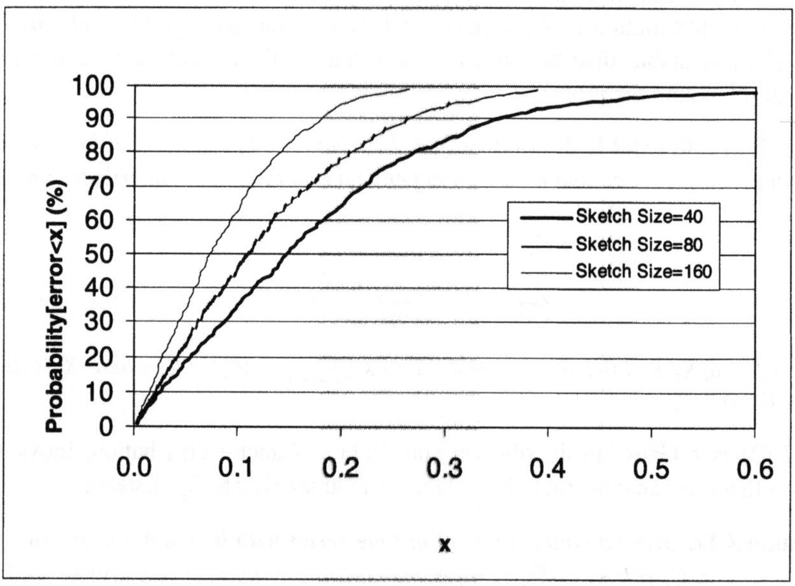

(b) Price return time series

Figure 2.20. Approximation of distances between time series using sketches; 2 hours of stock data.

In the above lemma, $|\mathbf{s}(\mathbf{x}) - \mathbf{s}(\mathbf{y})|$ is a vector with size k. The median of $|\mathbf{s}(\mathbf{x}) - \mathbf{s}(\mathbf{y})|$ is the median of the k values in the vector. It turns out that the scaling factor is 1 for both $p = 1$ (Cauchy distribution) and $p = 2$ (Gaussian distribution).

It is also possible to approximate the Hamming distance L_0 between pairs of time series using stable distributions. The reader can refer to recent results in [24].

In the time series data mining research, sketch-based approaches were used to identify representative trends [47, 25], to compute approximate wavelet coefficients [38], etc. Sketches have also many applications in streaming data management, including multidimensional histograms [89], data cleaning [28], and complex query processing [31, 27].

2.4.3 Sketches – The Bottom Line

If there are k sketches, then comparing a window of length n to those sketches yields a vector with only k elements each of which is a distance between that window and one sketch. Assuming $k << n$, this gives a substantial reduction in the size of the representation. When comparing two such time series windows, we can compare the sketches alone and obtain high accuracy. This reduces the time for inner product from $O(n)$ to $O(k)$. Even better, if k is small enough and we are interested only in highly correlated pairs (i.e., pairs with a large inner product), we may be able to use a multi-dimensional index structure, so most pairs need not even be compared. Even when k is large (over 10), we can partition the k-length vectors into smaller vectors and create a multi-dimensional index for each sub-vector. Then we can say that the two k-length vectors are close if a large fraction of those sub-vectors are close. The size of this fraction depends on the application and the degree of closeness required. The fraction can be determined empirically. Remember that because the distances computed from sketches are approximate, we need to do some post-processing after the highly correlated pairs are reported.

2.5 Comparison of Data Reduction Techniques

Having discussed the four different data reduction techniques, we can now compare them. This will help you as the data analyst choose the right data reduction technique for any problem you face. The criteria for comparison are summarized in Table 2.4.

First we discuss the time complexity to perform the data reduction for a time series having length n.

- The time complexity to compute the Discrete Fourier Transform (DFT) is $O(n^2)$ using the standard construction of the DFT. To compute the first k DFT coefficients will take $O(kn)$ time however. Also, the complexity of DFT can be reduced to $O(n \log n)$ using the Fast Fourier Transform algorithm. Therefore, to compute the first k DFT coefficients will take $\min\left(O(n \log n), O(kn)\right)$.
- The time complexity for a Discrete Wavelet Transform (DWT) computation is lower, $O(n)$.

- The time complexity of the singular value decomposition (SVD) depends on the size of the collection of time series under consideration. For a collection of $m, m \gg n$ time series, SVD takes time $O(m + n^3)$. The SVD for each time series requires $O(\frac{m}{n} + n^2)$ time. This is the slowest among all the data reduction techniques we have discussed.
- The time complexity for the random projection computation is $O(nk)$, where k is the number of sketches.

DFT, DWT and SVD are all based on orthogonal transforms. From the coefficients of the data reduction, we can reconstruct an approximation of the time series. By comparison, random projection is not based on any orthogonal transform. We cannot reconstruct an approximation of the time series. Fortunately, matching one time series against the other does not require that our synopsis technique gives a good approximation of the original.

In terms of distance approximation, DFT, DWT and SVD can be used for the approximation of only Euclidean (L_2) distance with one exception. Piecewise Aggregate Approximation (PAA), a transform closely related to the Discrete Haar Wavelet Transform, can handle any distance metric L_p, $p \neq 2$.

The next criterion has to do with the methods for constructing basis vectors for each of these data reduction technique. For the DFT, the basis vectors are fixed to be vectors based on trigonometric functions. One particular benefit of using DWT is that one can choose from a vast number of wavelet families of basis vectors. SVD is desirable in many cases because the basis vectors are data dependent. These vectors are computed from the data to achieve minimal approximation error. But this also implies that we need to store the basis vectors in addition to the SVD coefficients if we want to reconstruct the time series. The basis vectors of the random projection are chosen, well, randomly.

To approximate a time series by a few coefficients, the DFT, DWT and SVD all require the existence of some principal components in the time series data. Random projection, by contrast, does not make any assumption about the data. It can work even for white noise. This makes random projection very desirable for time series data having no obvious trends such as price differences in stock market data.

A particular drawback of DFT as a data reduction method is that the basis vectors of DFT do not have compact support. This makes it very hard for DFT to approximate time series having short term bursts or jumps. Most of the DWT basis vectors have compact support. Therefore, DWT can approximate a time series with jumps, but we need to choose a subset of coefficients that may not necessarily be the first few coefficients. SVD deals with the problem of discontinuity in the time series data more gracefully. If short term bursts or jumps are observed at the same location of most time series, it will be reflected by the basis vectors of the SVD at that location.

Figure 2.21 presents a decision tree to help you choose the right data reduction technique given the characteristics of your time series data.

Finally, in Figure 2.22 we remind you of the general approach we follow when applying a transform to the problem of finding highly correlated pairs of time series.

Table 2.4. Comparison of data reduction techniques.

Data Reduction Technique	DFT	DWT	SVD	Random Projection
Time Complexity	$n \log n$	n	$\frac{m}{n} + n^2$	nk
Based on Orthogonal Transform	Yes	Yes	Yes	No
Approximation of Time Series	Yes	Yes	Yes	No
L_p Distance	$p = 2$	$p = 2$	$p = 2$	$p = [0, 2]$
Basis Vectors	fixed one choice	fixed many choices	adaptive optimal	random
Require Existence of Principal Components	Yes	Yes	Yes	No
Handles Bursts	No	Yes	Yes	Yes

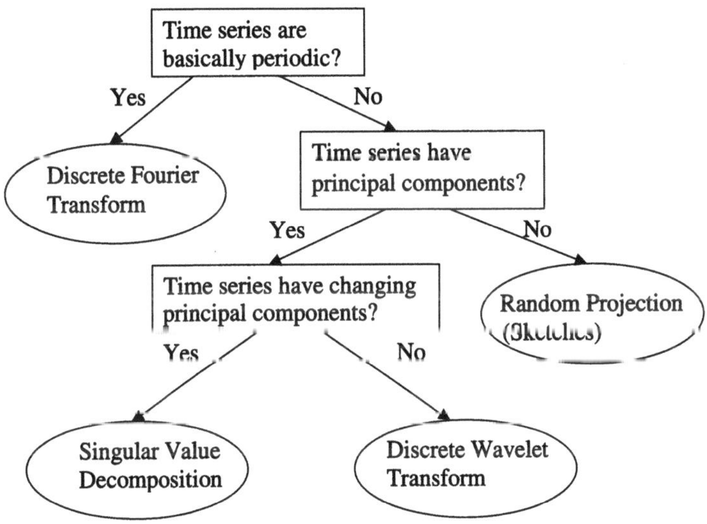

Figure 2.21. A decision tree for choosing the best data reduction technique.

This idea of transforming, reducing, computing, and then testing is central to all the techniques of this book.

2.6 Questions

1. a) Write a program in your favorite language (C, MATLAB, K, etc.) to implement the computation of Discrete Fourier Transform.
   b) Write a function that generates a random walk time series of length n.

```
Given: A collection of time series D
Return: All pairs of time series x^n, y^n ∈ D such that
    Corr(x^n, y^n) > 1 − ε

FOR EACH s^n ∈ D
  ŝ^n= NormalForm (s^n);
  S^N  = Transform(ŝ^n);//N ≪ n
ENDFOR
FOR EACH pair x^n, y^n ∈ D
  IF D^2(X^N, Y^N) < 2nε (in the transformed space)
    Retrieve the original time series x^n and y^n
    IF Corr(x^n, y^n) > 1 − ε (check in real space)
      Report that x^n, y^n are highly correlated.
    ENDIF
  ENDIF
ENDFOR
```

Figure 2.22. Algorithm to search for high correlation using your favorite transform.

   c) Compute the Discrete Fourier Transform of the random walk time series you generate.
2. Prove theorem 2.12(b).
3. a) Download the Fast Fourier Transform program from http://www.fftw.org/ and perform the FFT on the same time series you generate. Does it give you the same results as your codes?
   b) Optimize your codes to make your DFT routine as fast as possible, compare the speed of your routine and the FFT routine with different n. What conclusion do you get from the experiments?
4. Using the Daubechies-2 wavelet filters in (2.111) to compute the Discrete Wavelet Transform of the following time series:

$$1, 3, 5, 7, 4, 1, 5, 12, 17.$$

   Also compute the Inverse Discrete Wavelet Transform on the transformed time series you get. Do you get back the original time series?
5. a) Generate a small dataset of 200 random walk time series, each with length 128.
   b) Using a standard Singular Value Decomposition routine (such as that from LAPACK, MATLAB), compute the SVD on the time series dataset. If you keep only the first 10 largest eigenvalues, and reconstruct the time series from these 10 principal components, what is the total sum of the approximation errors for all the time series?
   c) Do you have a faster way to compute the sum of approximation errors?
6. Given a time series $\mathbf{x} = (x_1, x_2, \ldots, x_N)$ and a sketch pool as in (2.142), we want to compute the sketches in a sliding window fashion. That is, we want to

compute the sketches of the following time series:

$$\mathbf{x}_1 = (x_1, x_2, \ldots, x_n),$$
$$\mathbf{x}_2 = (x_2, x_3, \ldots, x_{n+1}),$$
$$\ldots,$$
$$\mathbf{x}_{N-n+1} = (x_{N-n+1}, x_{N-n+2}, \ldots, x_N).$$

Can you design a fast algorithm for this? What is the time complexity?

7. a) If you approximate the time series dataset in question 5 using the Discrete Fourier Transform, what is the sum of approximation errors?
   b) What if you use the Discrete Haar Wavelet Transform?
   c) What conclusion do you get from the experiments?

3

Indexing Methods

For our purposes, an index is a data structure that allows fast retrieval of time series that are close to a query time series. In index terms, because a time series is represented as a point in some space, our searches are similarity point searches in a single or multidimensional space. Of course, indexes have been extremely well studied and your favorite database text should describe B-trees and at least some multidimensional structures in great depth. Our intent is to focus on those indexes that are most useful to time series analysis as well as how to use them.

We will start with the simplest and most frequently used index structure, the B-tree, in Section 3.1. A B-tree is a one-dimensional index structure. We next present a simple multidimensional index structure, the KD-B-tree in Section 3.2 and discuss a more advanced structure the R-tree and its friends in Section 3.3. Finally in Section 3.4 we discuss a simple yet effective multidimensional index structure, the grid file.

Because time series data is high dimensional and the performance of multi dimensional indexes deteriorates rapidly as the number of dimensions increases, the basic way to search in a high dimensional space is to search in several low dimensional spaces and then combine the answers. Exactly how to do this requires some experimentation as we point out in Section 3.5.

The chapter is an intentionally brief primer because the state of the art in both theory and implementation is advancing so rapidly that your major decision point will be whether to use many single or a few multi-dimensional indexes, not which one to implement. Indeed, a poor implementation of a "better" structure may often give poorer performance than a better implementation of a poorer one as has been shown empirically [88].

3.1 B-tree

Suppose we want to look for a name in the phone book. The last names in the phone book are ordered alphabetically. Nobody will go through the phone book from page 1 to the end of book. Instead, we do a binary search on the phone book, jumping back and forth until we find the page containing the name.

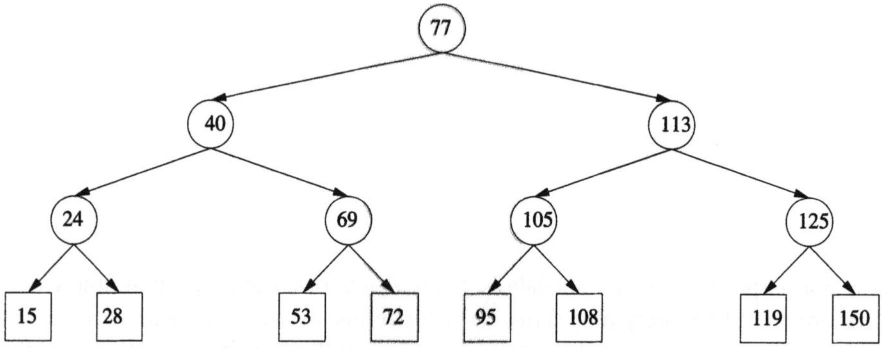

Figure 3.1. An example of a binary search tree.

A *binary search tree* is an index structure that allows this form of search for one-dimensional data. For example, if we have a collection of time series and we have already computed the average of each time series, then we might use a binary search tree to index the averages. Figure 3.1 shows an example of a binary search tree. Associated with each number in the node of the tree is a pointer to the time series so that we can retrieve the time series through the binary search tree. If we want to find a time series with 72 as the average, we first compared it to 77 as the root number. Because 72 is smaller than 77 we go to the left subtree of the root. 72 is compared to 40 and the right subtree is chosen. We reach 69 and take the right subtree of 69 to get to the leaf node 72.

The query above is a point query because we ask only for data with key equal to a specific value. A binary search tree can also answer range queries. A range query can be translated into two point queries. For example, to find all the time series with averages between 53 and 95, we first make a point query of 53 in the binary search tree and get the path p_1 from the root to the node 53. Similarly we find the path p_2 from the root to the node containing number 95. The region in the binary search tree between paths p_1 and p_2 contain all the data whose key is in the range $[53, 95]$.

The binary search tree in the above example is balanced in the sense that each leaf node is at the same distance from the root. If we have n indexed items and the binary search tree is balanced, the depth in the tree is $\log_2 n$. A search, either a point query or a range query, takes time $O(\log_2 n)$ using the binary search tree.

When the amount of data is large enough so that the binary search tree cannot fit into main memory, a binary search tree may have many levels in secondary memory. For example, if a search structure stores a trillion values, but the main memory can store only one billion, there will be ten levels of that structure on disk. This results from the low fanout of the binary search tree.

A *B-tree* generalizes the search strategy of a binary search tree by vastly increasing the *fanout* (to a thousand or more in modern systems). For the billion/trillion example, a B-tree in which each node has 1,000 children would require only one level in secondary memory, thus greatly reducing the disk access costs.

A B-tree is a balanced tree by construction, i.e., all leaves are at the same distance from the root node. As the example shows, the IO cost of a B-tree depends on the depth of the B-tree, because we have to reach the leaves of the B-tree to get the data. Therefore the shallower a B-tree, the fewer IO access a search costs. The depth of a B-tree is roughly $\log_{\text{fanout}} N$, where N is the number of data items.

Figure 3.2 shows an example of a B-tree. This B-tree has four levels and a fanout of 3. The leaves at the lowest level represent the data entries. Each non-leaf node contains a sequence of key-pointer pairs. The number of pairs is bounded by the fanout. The numbers shown in the non-leaf nodes are the keys. There is always a pointer associated with each key in the nodes. For key K_i, its associated pointer points to the subtree in which all key values are between K_i and K_{i+1}. If there are m keys, $K_1, K_2, \ldots, K_m$, in a node, there will be $m + 1$ pointers in that node, with P_0 pointing to a subtree in which all keys are less than K_1 and P_m pointing to a subtree in which all keys are larger than K_m. Also in most implementations, leaf nodes and non-leaf nodes at the same level are linked in a linked list to facilitate range queries.

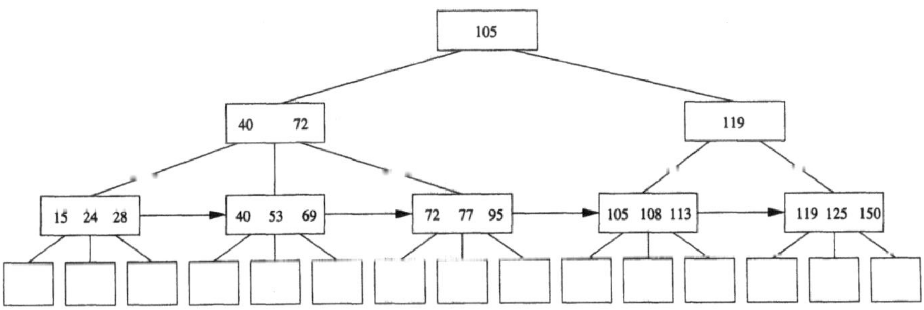

Figure 3.2. An example of a B-tree.

A query on a B-tree works in a similar fashion to a query on a binary search tree. Suppose we wish to find nodes with a search key value k. Starting at the root node, we identify the key K_i in the root node such that $K_i \leq k < K_{i+1}$. We follow P_i to the next node. If $k < K_1$, we follow pointer P_0. If $k > K_m$, we follow pointer P_m. The same process is repeated recursively until we get to the leaf node.

A B-tree is a dynamic index structure. We can insert and delete data in a B-tree without reconstructing the B-tree. When there are insertions and deletions, we must ensure the following:

1. There is no overflow. If the size of a node exceeds its allocated space after an insertion, the node must be split. A split in the child node will require an insertion of a key-pointer pair in the parent node. So the insertion might propagate upwards if the parent node overflows.

2. There is no underflow. If the size of a node becomes less than half full after a deletion, the node is in an underflow. We must merge this node with its sibling node in this case. This results in one less entry in the parent node. So the deletion might also propagate upwards. In practice, however, underflow is allowed. Research has shown [51] that merging a node (in fact freeing it) when it is empty is better than merging it when it falls below half full in most cases (specifically, in the common case that the B tree is growing in the average).
3. The B-tree remains balanced. We must adjust the B-tree after insertion and deletion to make sure that the B-tree is balanced. This is important because the search time would not be $O(\log n)$ if the B-tree is out of balance. Splits and merges preserve balance because they are propagated to the root.

3.2 KD-B-tree

Most queries on time series data are composed of multiple keys. For example, a query to find all the time series that start with values close to 10, 13, 15 can be thought of as a query with composite key (10, 13, 15) and a small range. To facilitate such similarity time series queries, we need indexing methods for high-dimensional spaces.

A *KD-tree* extends the binary search tree to higher dimensions. First let us consider the two-dimensional case. To build a KD-tree, we still split the data into two parts of roughly the same size based on the key. However, the keys we consider now are two-dimensional, that is, each is composed of two keys. Let us call the two keys the x and y coordinates, and the key is therefore a point in the xy plane.

A KD-tree will split the data based on the x and y coordinates alternately. For example, in Figure 3.3, there are 11 points in the xy plane. First we split these points by a line j that goes through one point, 3. This yields two subdivisions of the plane, one including all points whose x-coordinates are less than or equal to the x-coordinate of line j, the other including those with x-coordinates greater than the x-coordinate of line j. This next step is to split each subdivision horizontally. The left subdivision is separated by line c into two parts. The line c and above includes points 1, 2 and 3, and the subdivision below the line c includes points 4, 5, 6. This procedure is performed recursively on the subdivision. The splitting lines chosen cycle between vertical and horizontal, and all the lines go through a point in the split region.

The subdivision of the plane in Figure 3.3 corresponds directly to the KD-tree structure of Figure 3.4. The leaf nodes contain points while the non-leaf nodes contain lines. The non-leaf nodes at the even numbered levels correspond to vertical lines, while the non-leaf nodes at the odd numbered levels correspond to horizontal lines. The algorithm to construct a KD-tree is given in Figure 3.5

A search for point (K_x, K_y) on the KD-tree will start from the root. First we compare K_x to the line at level 0 to decide whether to go down to the left or right subtree. At the odd levels, K_y is compared with the key (a horizontal line) of the node. Similarly, at even levels, K_x is compared with the key (a vertical line) of the node. This is repeated until a leaf node is reached.

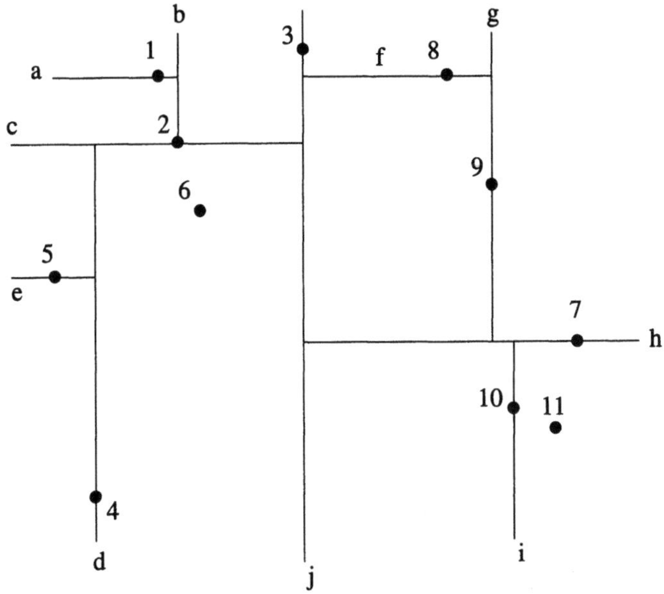

Figure 3.3. The subdivision of the plane with a KD-tree.

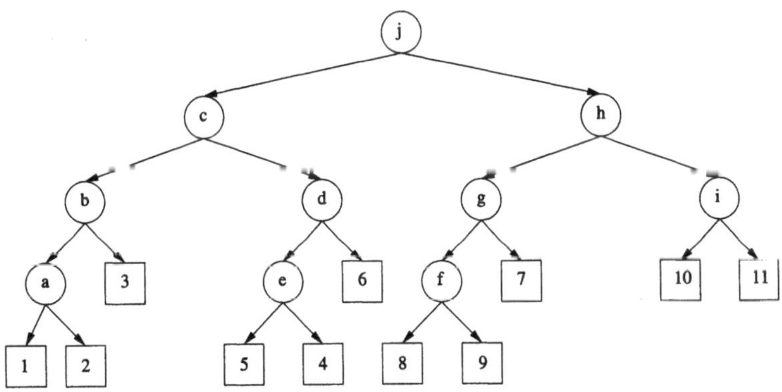

Figure 3.4. An example of a KD-tree.

Just as the B-tree is the secondary storage version of the binary search tree, the *KD-B-tree* is the secondary storage extension of the KD-tree. The KD-B-tree allows multiple child nodes for each non-leaf node to reduce the depth of the tree and guarantees balance. We omit the details here.

Another index structure for high dimensional data is the *quadtree*. A quadtree in two dimensions splits the plane in a different fashion than the KD-tree. At each level, a quadtree will split the plane at x and y coordinates of the midpoint simultaneously. This results in four rectangular regions of exactly the same size. An example of the

```
Algorithm BuildKDTree(P,d)
Given : A set of points P, the current depth d.
Return: A node in the KD-tree that is the root of a subtree
      containing P.

IF |P|=1
   RETURN a leaf that contains the only point
ELSE
   Create a node n
   IF d is even
      split P with a vertical line l through a point having
         the median x-coordinate.
      P1 are points with x-coordinates less than or equal to
         the x-coordinate of line l.
      P2 are points with x-coordinates greater than the
         x-coordinate of line l.
      Node n store the x-coordinate of l.
   ELSE  split P with a horizontal line l through a point
         with the median y-coordinate.
      P1 are points with y-coordinates greater than the
         y-coordinate of line l.
      P2 are points with x-coordinates less than or equal to
         the y-coordinate of line l.
      Node n stores the y-coordinate of l.
   ENDIF
   The left pointer of n is BuildKDTree(P1,d+1)
   The right pointer of n is BuildKDTree(P2,d+1)
   RETURN n
ENDIF
```

Figure 3.5. Algorithm to construct a KD-tree.

subdivision of the plane is shown in Figures 3.6 and 3.7. Similarly, a quadtree in k-dimensional space will split the space into 2^k hyper-boxes of equal size.

3.3 R-tree

The *R-tree* [43] extends the popular B-tree to higher dimensions. If it is well implemented, it is an efficient indexing structure for higher dimensional data, including points and regions.

Like the B-tree, an R-tree is a height-balanced tree with indexed data in the leaf nodes. In B-trees, each non-leaf node corresponds to an interval. Extended to higher dimensions, each non-leaf node in the R-tree corresponds to a bounding box (higher-dimensional intervals), called the *Minimum Bounding Boxes (MBB)*, in the indexed space. The MBB is the basic object of an R-tree. In a B-tree, the interval associated with a node includes all the intervals of its children nodes; in an R-tree, the bounding

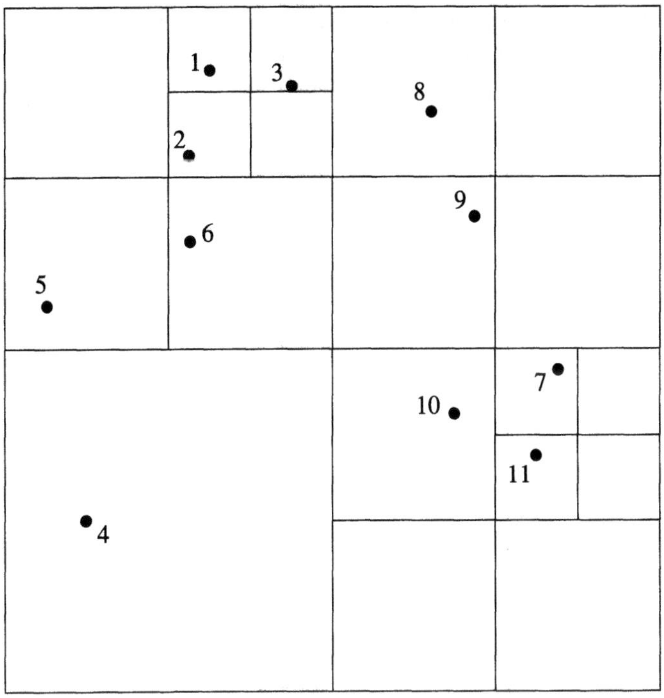

Figure 3.6. Subdividing the plane with a quadtree.

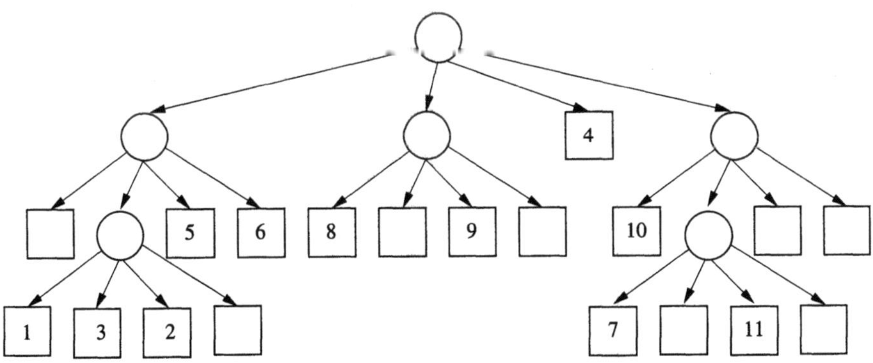

Figure 3.7. An example of a quadtree.

box associated with a node also includes all the bounding boxes of its children nodes. In a B-tree, the interval associated with a node will not overlap the intervals associated with its sibling nodes. Therefore the number of nodes to be accessed in a search on a B-tree is the depth of the B-tree. In an R-tree, however, the bounding boxes associated with a node could overlap the bounding boxes associated with its sibling nodes. Therefore a search through an R-tree could split into multiple searches.

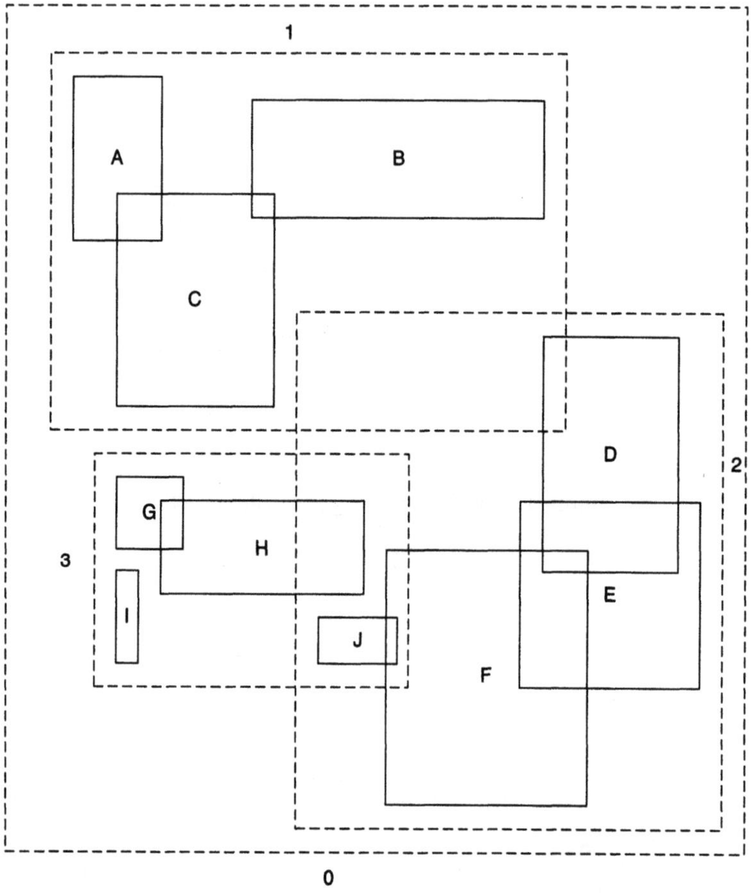

Figure 3.8. An example of the bounding boxes of an R-tree.

In Figure 3.8 we show an example of a set of rectangles and their bounding boxes. Though we show only a two-dimensional case for simplicity, the extension to higher dimensions is straightforward conceptually (though perhaps significantly slower). The corresponding organization of the R-tree is shown in Figure 3.9.

To search for all rectangles that contain a query point, we have to follow all children nodes with the bounding boxes containing the query point. Similarly, to search for all rectangles that intersect a query rectangle, we have to follow all children nodes with bounding boxes that intersect the query rectangle. A search in the R-tree could be slowed down because we have to follow multiple search paths.

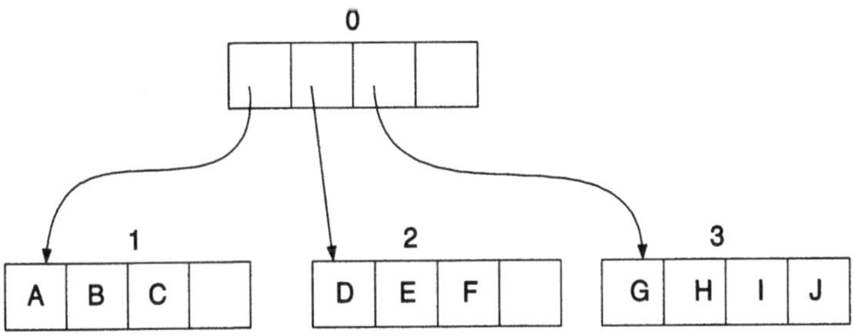

Figure 3.9. The R-tree structure.

To insert an object o into a R-tree, we first compute the MBB of the object $MBB(o)$. Insertions require the traversal of only a single path. When there are several candidate children nodes, the R-tree uses some heuristic algorithm to choose the best child node. Usually the criterion is that the bounding box of the chosen child node needs to be increased least. This will make the bounding box of the nodes more compact and thus minimize the overlapping area of sibling nodes' bounding boxes. If a node is full, it must split and that split may cascade upwards. Also the tree must be adjusted upwards to retain balance. In fact, the variants of R-tree differ mostly in the insertion algorithm. Such variants include the R^**-tree* [15] and R^+*-tree* [86].

The deletion of data in the R-tree is similar to that in the B-tree. We first search for the deleted item in the leaf node and then delete it. If the deletion of the item makes the bounding box of the leaf node smaller, we will update the size of the bounding box. This information is propagated upwards. We may also merge a node with its sibling nodes if its bounding box is reduced to below a certain size. This information is also propagated upwards if necessary.

3.4 Grid Structure

The R-tree discussed above and its variants are the most popular higher-dimensional indexing methods. However, the implementation of the R-tree is highly non-trivial and there are substantial costs in maintaining the R-tree dynamically. In this section, we give a quick review of an extremely simple indexing structure for indexing higher-dimensional data: the *Grid Structure*. The simplicity of the grid structure maintenance makes it attractive where response time is critical.

In the main memory grid structure, we superimpose a d-dimensional orthogonal regular grid on the indexed space. In practice, the indexed space is bounded, so without loss of generality, we assume each dimension is contained in [0, 1]. Let the

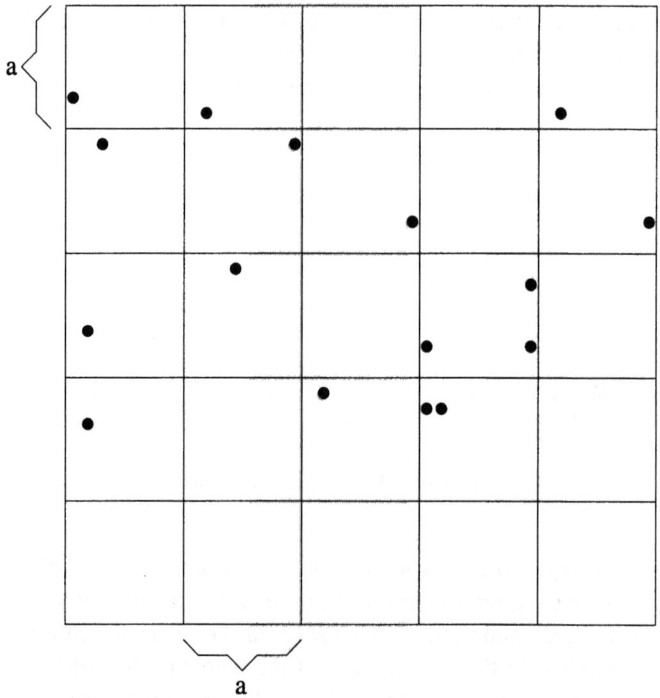

Figure 3.10. An example of a main memory grid structure.

spacing of the grid be a. Figure 3.10 shows such a two dimensional grid structure. We have partitioned the indexing space, a d-dimensional cube with diameter 1 into $\lceil \frac{1}{a} \rceil^d$ small cells. Each cell is a d-dimensional cube with side a. All the cells are stored in a d-dimensional array in the main memory.

In such a main memory grid structure, we can compute the cell a point belongs to by simple indexing. Let us use $(c_1, c_2, \ldots, c_d)$ to denote a cell that is the c_1-th in the first dimension and the c_2-th in the second dimension, etc. A point p with coordinates $x_1, x_2, \ldots, x_d$ is within the cell $(\frac{x_1}{a}, \frac{x_2}{a}, \ldots, \frac{x_d}{a})$. We say that point p is hashed to that cell.

With the help of the grid structure, we can find the point and its neighbors fast. Given a query point p, we compute the cell c to which p is hashed. The data of the query point p can then be retrieved. Also the data of points close to p are also hashed to cells near c, and can be retrieved efficiently too.

Insertion and deletion in the grid structure entail almost no maintenance cost. To insert a point p, we just compute the cell c that p is hashed to and append the data of p into the cell. To delete a point p, we just compute the cell c and delete the data of p from the cell.

Of course, the grid structure is not necessarily uniform. The partitioning of the indexed space can be different in different dimensions to adapt to the data and to reduce the number of cells in the grid structure.

The grid structure is well suited for point queries. Query and update in a grid structure is faster than in other high-dimensional index structures. However, the space requirement for grid structures is very high. There will be $\lceil \frac{1}{a} \rceil^d$ cells in a d-dimensional space. So the grid structure is effective only when indexing lower dimensional spaces that are reasonably uniform.

The *Grid File* is a memory index structure based on the grid structure. The goal of the grid file to guarantee that any access of data requires at most two IO operations.

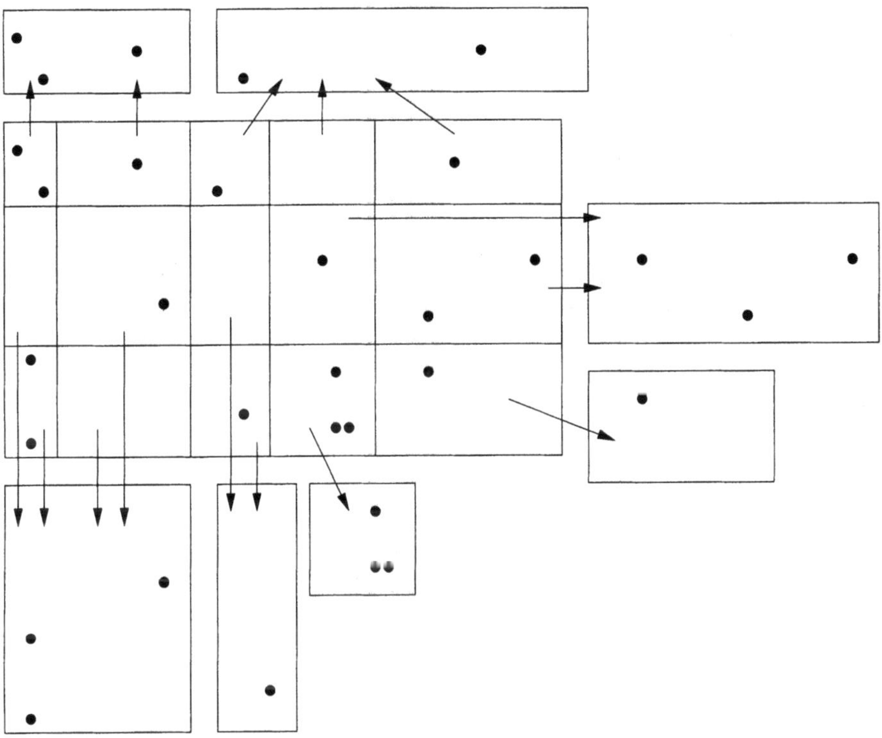

Figure 3.11. An example of a grid file structure.

Figure 3.11 shows an example of a grid file. In a grid file, the partition of the indexed space is unequal in the different dimensions. The grid is kept in main memory. Each cell in the grid is associated with a grid block. This is shown as a pointer in the figure. Each grid cell can be associated with only one grid block, but there can be many grid cells associated with one grid block. A grid block is made up of one or several grid cells, as long as the union of these grid cells is a high-dimensional box. There can be only up to m data points in a grid block. For purposes of illustration, m is 3 in this example. The constraint on the number of points in a grid block is to guarantee that the grid block can fit into a page in secondary memory.

To search for a data point in high dimension, we first locate which cell the query point is hashed to based on the grid structure. If the grid cell is not in main memory, we perform one disk access to retrieve the grid block that contains the cell. From the loaded cell, we can access the page that contains the data associated with the query point. Therefore, any point query in a grid file requires at most two IO accesses.

Insertion and deletion in a grid file is more complicated than in a grid structure. When a point p is inserted, we find the grid cell that p hashes to and the grid block that contains the cell. If the grid block exceeds its capacity, the grid block must be split. Because the grid file requires that the grid block be a high-dimensional box, a split of the grid block is not a local operation. All the other grid blocks intersected by the splitting line (or plane) have to be updated too. This is similar in spirit to what an R-tree does.

Compared to other multidimensional index methods, the grid file is superior when the data is very dense. Also it is suitable when the data dimensionality is relatively low.

In summary, Figure 3.12 shows a decision tree at a high level, based on which we can choose the secondary storage index method for different data. Note that we use the data dimensionality of 4 to draw the line between a grid file and other multidimensional index structures. Please take this as rule of thumb rather than a mathematical fact.

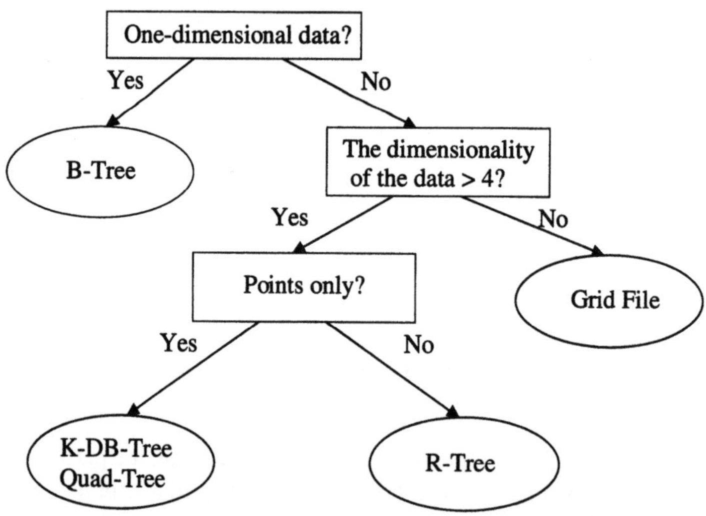

Figure 3.12. A decision tree for choosing an index method.

3.5 Indexes – The Bottom Line

Suppose you want to find the time series from a database D that are sufficiently close (within distance c) to a query time series q. Even after data reduction and transformation, however, the number of dimensions n may be too large for any multi-dimensional index to handle well. In that case, you'll want to combine the results from multiple indexes using the following framework:

1. Partition the dimensions into p subsets having m or fewer dimensions each where m is relatively small (under 8 is a good heuristic choice or 4 if you're not using a grid file).
2. Characterize the query and the time series in D based on this partitioning. Thus each time series in D is a point in p data structures.
3. Perform a similarity search for q in each of the p data structures.
4. If q is within distance c of some time series t in every one of the p data structures, then q and t may be within distance c of one another. (This may give many false positives, because the distance may be greater than c yet be within c for every p data structures. So this last step can be tuned. But this scheme is at least safe.)

3.6 Questions

1. How will the subdivision of the plane with a KD tree in Figure 3.3 change if we add a point in the middle of point 8 and 9?
2. Is the KD-tree a balanced tree? How about the quadtree?
3. Refer to the literature of R^*-tree [15] and summarize the major differences between an R-tree and an R^*-tree.
4. For a main memory grid structure, if the indexing space is a 10-dimensional cube with diameter 1 and each cell is a cube with diameter 0.02, how many cells do we need to store the structure in main memory? If each cell takes space 4 bytes, what is the size of the memory we need to allocate for the grid structure? How might hashing help?

4

Flexible Similarity Search

There are many applications for similarity search in time series data of which the following are just a small sample.

1. In finance, a trader may be interested in finding pairs of stocks that move similarly, perhaps with some lag.
2. In music, a person may want to find a song that is similar to one that he can hum.
3. In business management, spotting products with similar selling patterns can result in more efficient product management.
4. In environmental science, by comparing the pollutant level in different sections of a river, scientists can have a better understanding of environmental changes.

Humans are often good at detecting the similarity between time series by just looking at their plots of time versus value. Such knowledge must be encoded in the computer if we want to automate the detection of similarity among time series.

Formally, given a pair of time series, their similarity is measured by their correlation or distance. If we treat a time series as a high dimensional point, the Euclidean distance would seem to be a natural measure for the distance between time series.

The Euclidean distance measure is not adequately flexible to be a similarity measure between time series however for the following reasons:

1. Two time series can be very similar even though they have different base lines or amplitude scales. For example, the price movements of two stocks that follow the same pattern might have a large Euclidean distance between them because they are moving around different baseline prices. (Whereas correlation overcomes this objection, because it effectively normalizes the distance, it does not adequately address the next ones.)
2. The Euclidean distance between two time series of different lengths is undefined even though the time series are similar to each other. For example, a musical piece sounds similar even when it is played at slightly different tempos, but then the length of the two time series will be different.
3. Two time series could be very similar even though they are not perfectly synchronized. The Euclidean distance that sums up the difference between each pair of

corresponding data items between two time series is too rigid and will amplify the difference between time series.

Because the Euclidean distance alone is too rigid, a flexible similarity measure must incorporate manipulations of time series. This chapter discusses those manipulations.

As background, we will discuss a general framework for indexing time series databases based on the Euclidean distance. Even though we've maligned that distance measure, the Euclidean distance remains the basis for other time series similarity measures and is intuitive.

This framework will build on the data reduction and manipulation techniques we discussed in Chapter 2 and the indexing methods of Chapter 3. Next we will discuss how to allow time series to be compared even though they have different amplitude baselines and scales in Section 4.2. Section 4.3 discusses how to compensate for different sampling rates. We will also discuss manipulations that take into consideration local deviations in the synchronization between time series in Section 4.4.

4.1 GEMINI Framework

The GEMINI framework [8] handles two categories of Euclidean similarity queries in time series databases.

1. **Whole Sequence Matching:** In whole sequence matching, all the time series in the database are of the same length n. The query time series q is of length n too. The Euclidean distance between the query time series and any time series in the database can be computed in linear time. Given a query threshold ϵ, the answer to a whole sequence similarity query for q are all the time series in the database whose Euclidean distance with q are less than the threshold ϵ.
2. **Subsequence Matching:** In subsequence matching, the time series in the database can have different lengths. The lengths of these candidate time series are usually larger than the length of the query time series. The answer to a subsequence query is any subsequence of any candidate time series whose distance with q is less than ϵ.

A naive way to tackle the problem of similarity querying in time series databases is *linear scan*. In linear scan, one computes the Euclidean distance between the query time series and all the candidate time series (all subsequences of the candidate time series for subsequence matching) in the database. Those time series having distances less than ϵ to the query are reported.

Linear scan scales poorly because we have to read all the time series in the database. Therefore the computing time increases linearly with the size of the database.

Because a time series of length n can be seen as a point in an n-dimensional space, we can index all the time series using an n-dimensional index structure. A similarity search for a query time series q is just a range query of the query point in n-dimensional space.

Unfortunately, the above indexing method is impractical. All multidimensional index methods suffer from the "curse of dimensionality." That is, as the dimensionality of the index structure increases, the performance of the index structures deteriorates. For example, an R+ tree can usefully index a space having dimensionality up to at most 5 to 10.

In their seminal work [8], Agrawal, Faloutsos and Swami investigate the problem of how to index the time series database for whole sequence matching. To achieve similarity search in time series with high performance, they introduce the GEMINI framework [8, 34]. It was extended to subsequence matching in follow-up research [34].

Similarity search in the GEMINI framework is based on dimensionality reduction. The difficulty of indexing time series comes from the high dimensionality of the data. Indexing time series within the GEMINI framework will use data reduction techniques to reduce the dimensionality of time series. This results in a concise representation of time series in a lower dimension, which is also known as the *feature space*.

Formally, given a time series $\mathbf{x}^n$, a dimensionality reduction transform $\mathcal{T}$ will reduce it to a lower dimension $\mathbf{X}^k = \mathcal{T}(\mathbf{x}^n), k \ll n$. $\mathbf{X}^k$ is also called the *feature vector* or *signature* of $\mathbf{x}^n$. After the time series are mapped to a lower dimensional space, they can perhaps be indexed by a multidimensional index structure such as an R*-tree or a grid file. This was part of the motivation for reduction we presented in Chapter 2.

Because the feature vector is an approximation of the original time series, a query on the indexed feature space can get only approximate answers. There are two kinds of approximation errors for a similarity query.

1. *False Negatives* The approximate query answers do not include some time series in the database that are actually qualified answers.
2. *False Positives* The approximate query answers include some time series in the database that are actually not qualified answers.

False Negatives affect correctness whereas False Positives affect only performance. Correctness is more important than performance for most applications.

Next we will prove an important lemma in time series data mining. It says that if the dimensionality reduction transform $\mathcal{T}$ has some special property, we can guarantee there are no false negatives in the approximate query answer.

Lemma 4.1 (Lower Bound). [34] *To guarantee no false negatives in similarity queries, $\mathcal{T}$ must be lower-bounding, that is, the distance between time series under dimensionality reduction should lower-bound their original distance:*

$$D(\mathcal{T}(\mathbf{x}), \mathcal{T}(\mathbf{y})) \le D(\mathbf{x}, \mathbf{y}). \tag{4.1}$$

Proof. Suppose that the query range is ϵ. Let q be the query time series, and x a qualifying time series in the database. Also suppose that their feature vectors are $X = \mathcal{T}(x)$ and $Q = \mathcal{T}(q)$ respectively.

To guarantee no false negatives, any qualifying time series must be retrieved by the index structure in the feature space. In other words,

$$\text{If} \quad D(q,x) \le \epsilon \quad \text{then} \quad D(Q,X) \le \epsilon. \tag{4.2}$$

From (4.1), we know that

$$D(Q,X) = D(\mathcal{T}(q), \mathcal{T}(x)) \le D(q,x) \le \epsilon.$$

Therefore, there are no false negatives in the result. □

Most popular dimensionality reduction transformations, including the Fourier Transform, the Wavelet Transform, the Singular Value Decomposition and the Piecewise Aggregate Approximation (discussed below), are lower-bounding for Euclidean distance. Random projection can approximate the Euclidean distance with a probabilistic guaranteed bound. Therefore random projection may suffer some false negatives.

In the first stage of similarity query, we retrieve similar time series based on the indexed feature space. We still have some false positives in our answer set. These false positives can be further pruned away by examining the result set returned in the first stage. In this post-processing stage, we will retrieve the original time series instead of the feature vectors, and compute the true distance between the query time series and the candidate time series, thus eliminating false positives. The post-processing stage will perform a linear scan on all the time series returned in the first stage. This is fast in practice because the computation of the first stage already prunes lots of time series based on the feature vectors.

For subsequence queries, we can impose sliding windows of fixed length n on a long time series in the database. Each subsequence is mapped to its feature vector, a point in the feature space. Therefore, a long time series is mapped to a trail of points in the feature space. These trails can be further divided into sub-trails based on their proximity. The minimum bounding rectangles of the sub-trails in the feather space can be indexed by multidimensional indexes. The answers to a subsequence query are included in all those sub-trails with minimum bounding rectangles overlapping with the query rectangle. For more details, the readers can refer to [34].

4.2 Shifting and Scaling

As we mentioned before, the Euclidean distance alone is insufficiently flexible. Consider the following example.

In Figure 4.1, we show the stock price movements of IBM, LXK (Lexmark) and MMM (3M COMPANY) in the year 2000. From the time series plots, we can see that the price movements of IBM are much more similar to that of LXK than to MMM. However, the Euclidean distance between the time series of IBM and LXK is 9.2×10^5, which is much greater than that between IBM and MMM (8.2×10^4).

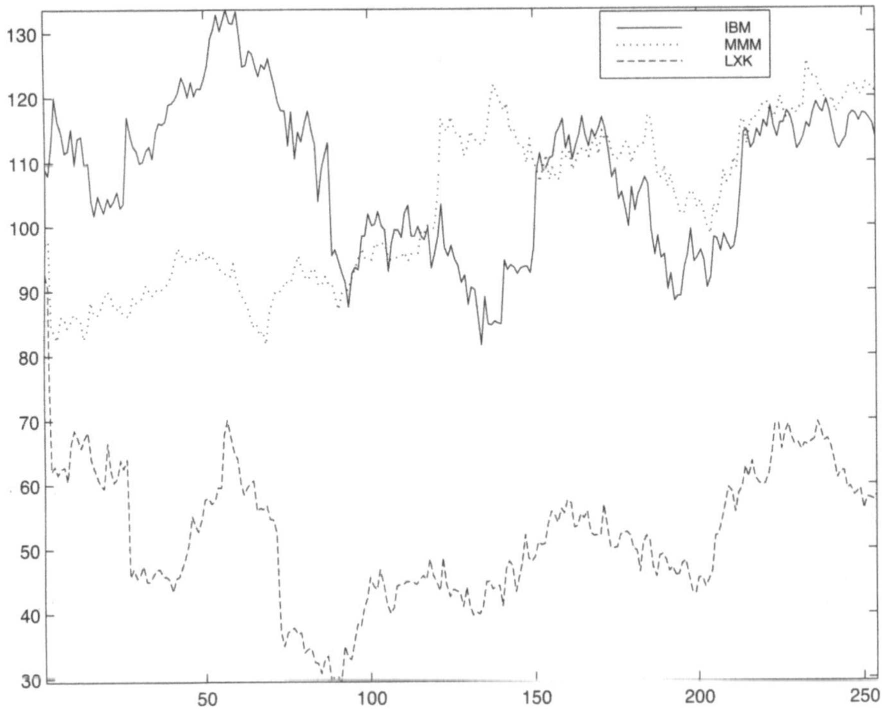

Figure 4.1. Stock price time series of IBM, LXK and MMM in year 2000.

From the above example, we can see that a good similarity measure should allow the time series to be shifted and scaled before we measure their Euclidean distance. Here we define the shifting and scaling transforms.

The *Shifting* transform on a time series yields a new time series by adding some real number to each item in the original time series.

Definition 4.2 (Shifting). *A shifting by* δ *on a time series* $\mathbf{x} = \big(x(1), x(2), \ldots, x(n)\big)$ *is* $\mathbf{x} + \delta = \big(x(1) + \delta, x(2) + \delta, \ldots, x(n) + \delta\big)$.

The *Scaling* transform on a time series yields a new time series by multiplying some real number to each item in the original time series.

Definition 4.3 (Scaling). *A scaling of* b *on a time series* $\mathbf{x} = \big(x(1), x(2), \ldots, x(n)\big)$ *is* $b\mathbf{x} = \big(bx(1), bx(2), \ldots, bx(n)\big)$.

A simple way to make a similarity measure invariant to shifting and scaling is to normalize the time series first.

Definition 4.4 (Normal Form). *The normal form* $\hat{\mathbf{x}}$ *of a time series* $\mathbf{x}$ *is transformed from* $\mathbf{x}$ *by shifting the time series by its mean and then scaling by its standard deviation.*

$$\hat{\mathbf{x}} = \frac{\mathbf{x} - \text{avg}(\mathbf{x})}{\text{std}(\mathbf{x})} \tag{4.3}$$

It is obvious that the normal form of a time series has the following properties.

Lemma 4.5. *Let the normal form of time series* $\mathbf{x}$ *be* $\hat{\mathbf{x}}$; *then*

$$\sum_{i=1}^{n} \hat{x}(i) = 0, \quad (\text{avg}(\hat{\mathbf{x}}) = 0) \tag{4.4}$$

$$\sum_{i=1}^{n} \hat{x}^2(i) = n \quad (\text{std}(\hat{\mathbf{x}}) = \sqrt{n}). \tag{4.5}$$

The Euclidean distance between the normal forms of two time series is a similarity measure between time series that is invariant to shifting and scaling, because they have the same baseline and scale. For example, Figure 4.2 shows the time series of the same stock as in Figure 4.2, except they are normalized. We can see that the Euclidean distance between the normal forms of the series of IBM and LXK is smaller than that between IBM and MMM.

The *norm* of a normal form time series of length n is n. Therefore the Euclidean distance between the normal forms of time series ranges between 0 and $2n$.

It is well known that the Pearson Correlation Coefficient is a similarity measure that is also invariant to shifting and scaling. To see this, recall the definition of *Pearson Correlation Coefficient.*

Definition 4.6 (Pearson Correlation Coefficient). *The Pearson Correlation Coefficient between two time series is defined as*

$$\text{corr}(\mathbf{x}, \mathbf{y}) = \frac{\text{avg}(\mathbf{x} * \mathbf{y}) - \text{avg}(\mathbf{x})\,\text{avg}(\mathbf{y})}{\text{std}(\mathbf{x})\,\text{std}(\mathbf{y})} \tag{4.6}$$

The Pearson Correlation Coefficient ranges from -1 to 1. It is not surprising that the Pearson Correlation Coefficient is closely related to the Euclidean distance between normal forms.

Lemma 4.7.

$$D^2(\hat{\mathbf{x}}, \hat{\mathbf{y}}) = 2n\big(1 - \text{corr}(\mathbf{x}, \mathbf{y})\big). \tag{4.7}$$

Proof. The Pearson Correlation Coefficient can also be written as follows:

$$\begin{aligned}
\text{corr}(\mathbf{x}, \mathbf{y}) &= \frac{\frac{1}{n}\sum_{i=1}^{n}\big(x(i)y(i)\big) - \text{avg}(\mathbf{x})\,\text{avg}(\mathbf{y})}{\text{std}(\mathbf{x})\,\text{std}(\mathbf{y})} \\
&= \frac{\frac{1}{n}\sum_{i=1}^{n}\big(x(i) - \text{avg}(\mathbf{x})\big)\big(y(i) - \text{avg}(\mathbf{x})\big)}{\text{std}(\mathbf{x})\,\text{std}(\mathbf{y})} \\
&= \frac{1}{n}\sum_{i=1}^{n}\hat{x}(i)\hat{y}(i).
\end{aligned}$$

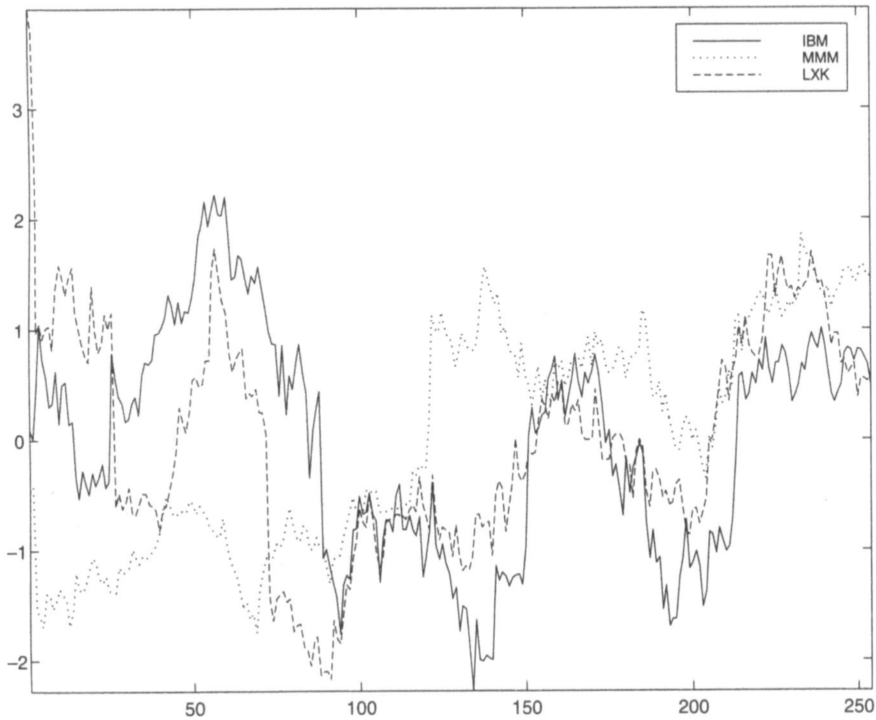

Figure 4.2. Normalized stock price time series of IBM, LXK and MMM in year 2000.

Therefore, we have

$$\begin{aligned} D^2(\hat{\mathbf{x}}, \hat{\mathbf{y}}) &= \sum_{i=1}^{n} (\hat{x}_i - \hat{y}_i)^2 \\ &= \sum_{i=1}^{n} \hat{x}^2(i) + \sum_{i=1}^{n} \hat{y}^2(i) - 2\sum_{i=1}^{n} \hat{x}(i)\hat{y}(i) \\ &= n + n - 2n\,\mathrm{corr}(\mathbf{x}, \mathbf{y}) \\ &= 2n\big(1 - \mathrm{corr}(\mathbf{x}, \mathbf{y})\big). \end{aligned}$$ □

The Euclidean distance between the normal forms of two time series is inversely related to their correlation. Small normalized Euclidean distances correspond to a higher correlation and vice-versa. Zero normalized Euclidean distance corresponds to a correlation of 1. A normalized Euclidean distance of n corresponds to 0 correlation. Also normalized Euclidean distances larger than n give negative correlation, with normalized Euclidean distances of $2n$ corresponding to the perfect anti-correlation -1.

In summary, to index time series for a similarity measure that is invariant to scaling and shifting, we need only to index the feature vectors for the normal forms of the time series. Given a query sequence, we also compute its feature vector of normal forms and then search with the indexes.

4.3 Time Scaling

The shifting and scaling transformations of the last section are performed on the amplitude axis of the time series. Similar transformations can be performed on the time axis.

Intuitively, time scaling stretches or squeezes the time axis of a sequence uniformly. This is also known as *Uniform Time Warping*. Given two time series of different lengths, we can compute their distance by time-scaling these two time series to a common length before performing a normalized distance comparison.

For example, in Figure 4.3 we show Dollar/Euro exchange rate time series between 1999 and 2002. The time series in Figure 4.3(a) is the daily exchange rate series and the series in Figure 4.3(b) is the monthly exchange rate series. Of course these two time series are very similar, because they record the same sequence of events at different sampling rates. We can see that if we stretch the monthly rate time series by a factor of approximately 30, or if we squeeze the daily rate series by a factor of approximately 30, we can make the two sequences have the same length. We can then compute their Euclidean distance.

The squeezing and stretching of the time axis can be formally defined as the upsample and the downsample transforms, respectively.[1] Some applications may upsample and downsample differently depending on the semantics of the time series. For example, weekly sales would have to be divided by the number of sale days to get daily sales, whereas stock prices could be upsampled in the way we say here or by an interpolating spline.

Definition 4.8 (w-upsampling). *The w-upsampling of a time series $\mathbf{x}^n$ is*

$$U^w(\mathbf{x}^n) = \mathbf{z}^{nw},$$

where

$$z_i = x_{\lfloor i/w \rfloor}, i = 0, 1, \ldots, nw - 1.$$

Definition 4.9 (w-downsampling). *The w-downsampling of a time series $\mathbf{x}^n$ is*

$$D^w(\mathbf{x}^n) = \mathbf{z}^{\lfloor n/w \rfloor},$$

where

$$z_i = x_{iw}, i = 0, 1, \ldots, \lfloor n/w \rfloor - 1.$$

[1] Note that here the definition is slightly different from those that we used for the Discrete Wavelet Transform in Chapter 2.

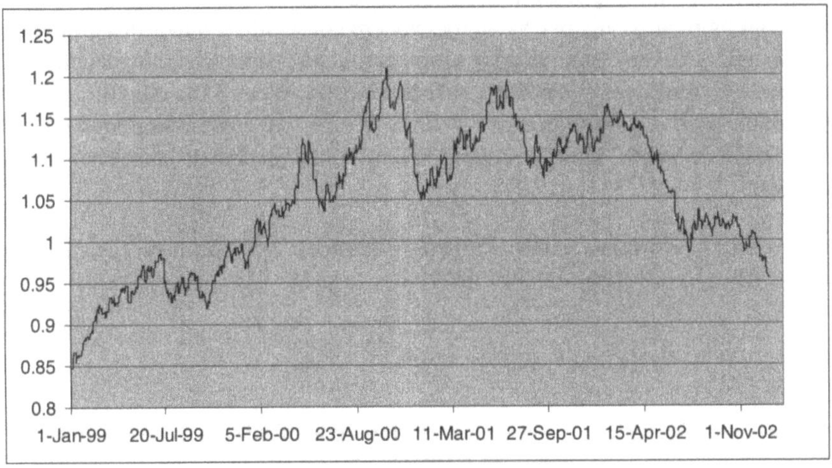

(a) Daily exchange rates of Dollar/Euro

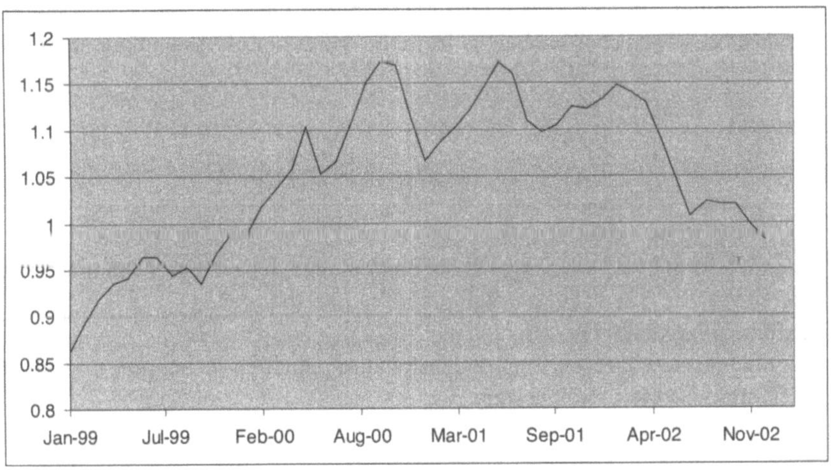

(b) Monthly exchange rates of Dollar/Euro

Figure 4.3. The Dollar/Euro exchange rate time series for different time scales.

Intuitively, w-upsampling repeats each value in a time series w times, while w-downsampling retains only one value for every m consecutive values in a time series.

Upsampling or downsampling by an integer factor works when the length of one time series is a multiple of that of another time series. What if the ratio of the length of two time series is a fraction, say 1.5? In such case, we can upsampling both time series to a larger length. This gives the Uniform Time Warping distance between two time series.

Definition 4.10 (Uniform Time Warping distance). *The Uniform Time Warping distance between two time series* $\mathbf{x}^n, \mathbf{y}^m$ *is*

$$D^2_{UTW}(\mathbf{x}^n, \mathbf{y}^m) = \frac{\sum_{i=1}^{mn-1} (x_{\lfloor i/m \rfloor} - y_{\lfloor i/n \rfloor})^2}{mn}. \tag{4.8}$$

For simplicity of notation, we stretch both time axis of $\mathbf{x}^n, \mathbf{y}^m$ to be mn. It is clear that if their greatest common divisor, $\mathrm{GCD}(m, n) > 1$, the time axis can be stretched to their least common multiple $\mathrm{LCM}(n, m)$. Also, any integer less than $\mathrm{LCM}(n, m)$ but much larger than $\max(m, n)$ can replace mn in (4.8) to get an approximation of the UTW distance.

The following lemma makes it possible to reduce the UTW distance to the Euclidean distance. The proof is obvious from the definition of upsampling and UTW distance.

Lemma 4.11.

$$D^2_{UTW}(\mathbf{x}^n, \mathbf{y}^m) = \frac{D^2\big(U^m(\mathbf{x}^n), U^n(\mathbf{y}^m)\big)}{mn}. \tag{4.9}$$

Using Uniform Time Warping, we can compute the distance between time series of different lengths. The UTW normal form of a time series $\mathbf{x}^n$ is $U^w(\mathbf{x}^n)$, where nw is a predefined large number.

It is not hard to adjust existing dimensionality reduction techniques to compute the UTW normal form of a time series.

4.4 Local Dynamic Time Warping

The standard definition of Dynamic Time Warping distance is as follows [17, 100, 77]:

Definition 4.12 (Dynamic Time Warping distance). *The Dynamic Time Warping distance between two time series* $\mathbf{x}, \mathbf{y}$ *is*

$$D^2_{DTW}(\mathbf{x}, \mathbf{y}) = D^2(\mathrm{First}(\mathbf{x}), \mathrm{First}(\mathbf{y})) + \min \begin{cases} D^2_{DTW}(\mathbf{x}, \mathrm{Rest}(\mathbf{y})) \\ D^2_{DTW}(\mathrm{Rest}(\mathbf{x}), \mathbf{y}) \\ D^2_{DTW}(\mathrm{Rest}(\mathbf{x}), \mathrm{Rest}(\mathbf{y})). \end{cases} \tag{4.10}$$

Intuitively, the Euclidean distance is the sum of synchronous point-by-point distances between two time series, while the Dynamic Time Warping distance allows the stretching or squeezing of the time axis to minimize the distance. This is illustrated in Figure 4.4.

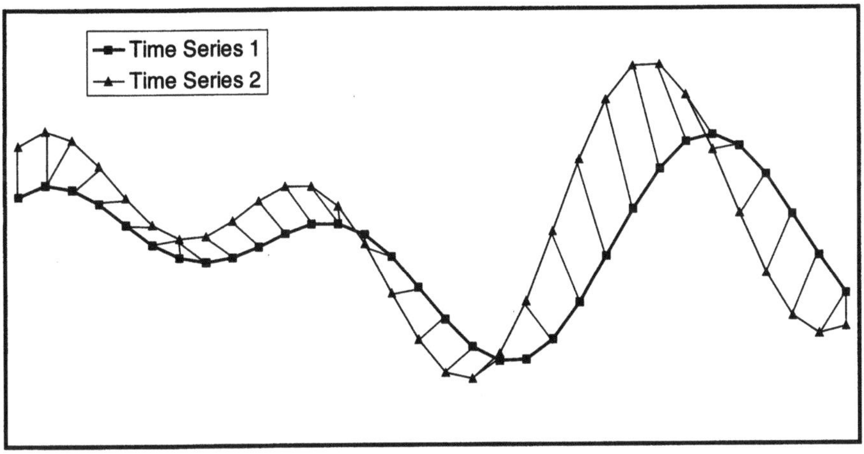

Figure 4.4. Illustrated computation of Dynamic Time Warping.

The process of computing the DTW distance can be visualized as a string matching-style dynamic program (Figure 4.5). We construct an $n \times m$ matrix to align time series $\mathbf{x}^n$ and $\mathbf{y}^m$. The cell (i, j) corresponds to the alignment of the element x_i with y_j. A warping path, P, from cell (0, 0) to $(n-1, m-1)$ corresponds to a particular alignment, element by element, between $\mathbf{x}^n$ and $\mathbf{y}^m$:

$$P = p_1, p_2, \ldots, p_L = (p_1^x, p_1^y), (p_2^x, p_2^y), \ldots, (p_L^x, p_L^y)$$

$$\max(n, m) \leq L \leq n + m - 1$$

$p_t^x, p_t^y t = 1, 2, \ldots, L$ are the position numbers of $\mathbf{x}^n$ and $\mathbf{y}^m$ respectively in the alignment. The distance between $\mathbf{x}^n$ and $\mathbf{y}^m$ on the warping path P is the distance between $x_{p_t^x}$ and $y_{p_t^y}, t = 1, 2, \ldots, L$. The constraints on the path P are

- P must be monotonic: $p_t^x - p_{t-1}^x \geq 0$ and $p_t^y - p_{t-1}^y \geq 0$
- P must be continuous: $p_t^x - p_{t-1}^x \leq 1$ and $p_t^y - p_{t-1}^y \leq 1$.

The number of possible warping paths grows exponentially with the length of the time series. The distance that is minimized over all paths is the Dynamic Time Warping distance. It can be computed using Dynamic Programming in $O(mn)$ time [17].

We can see that Uniform Time Warping (UTW) is a special case of Dynamic Time Warping. The constraint imposed by UTW is that the warping path must be diagonal.

An intermediate position between Uniform Time Warping and Dynamic Time Warping is Local Dynamic Time Warping (LDTW). Intuitively, humans will match two time series of different lengths as follows. First, the two time series are globally stretched to the same length. They are then compared locally point-by-point, with some warping within a small neighborhood in the time axis. Such a two-step transform can emulate traditional Dynamic Time Warping while avoiding some unintuitive results such as matching one point to almost the entire other sequence. Here is the definition of Local Dynamic Time Warping.

Definition 4.13 (k-Local Dynamic Time Warping distance). *The k-Local Dynamic Time Warping distance between two time series* **x**,**y** *is*

$$D^2_{\mathrm{LDTW}(k)}(\mathbf{x},\mathbf{y}) = D^2_{\mathrm{constraint}(k)}(\mathrm{First}(\mathbf{x}),\mathrm{First}(\mathbf{y}))$$

$$+\min\begin{cases} D^2_{\mathrm{LDTW}(k)}(\mathbf{x},\mathrm{Rest}(\mathbf{y})) \\ D^2_{\mathrm{LDTW}(k)}(\mathrm{Rest}(\mathbf{x}),\mathbf{y}) \\ D^2_{\mathrm{LDTW}(k)}(\mathrm{Rest}(\mathbf{x}),\mathrm{Rest}(\mathbf{y})) \end{cases} \tag{4.11}$$

$$D^2_{\mathrm{constraint}(k)}(x_i,y_j) = \begin{cases} D^2(x_i,y_j) \; \mathit{if}\; |i-j|\le k \\ \infty \qquad\qquad \mathit{if}\; |i-j|>k. \end{cases} \tag{4.12}$$

Figure 4.5 shows a warping path for $D_{\mathrm{LDTW}(2)}$in a time warping grid. The possible paths are constrained to be within the shadow, which is a beam of width $5(=2*2+1)$ along the diagonal path. The warping width is defined as $\delta = \frac{2k+1}{n}$. Such a constraint is also known as a Sakoe–Chiba Band. Other similar constraints are also discussed in [53]. It can be shown that the complexity of computing k-Local Dynamic Time Warping Distance is $O(kn)$ using dynamic programming.

Note that in the work [53], the definition of the DTW is actually LDTW. By combining UTW and LDTW together, we define a more general DTW distance:

Definition 4.14. *The Dynamic Time Warping distance between two time series is the LDTW distance between their UTW normal forms.*

In other words, it is the LDTW distance between two time series after they are both upsampled to be of the same length. In a slight abuse of notation, we will not distinguish between LDTW and DTW in the remaining chapters of the book, and we assume the distance of LDTW is computed after the UTW transform has been performed on the time series. In Chapter 6, we will show an indexing technique for time series databases allowing dynamic time warping.

In general, the choice of transformation for a similarity comparison between time series depends on the mechanism that generates the time series. The GEMINI framework can be extended to all of them. Different transformations can be combined. Figure 4.6 gives some advice for choosing among them.

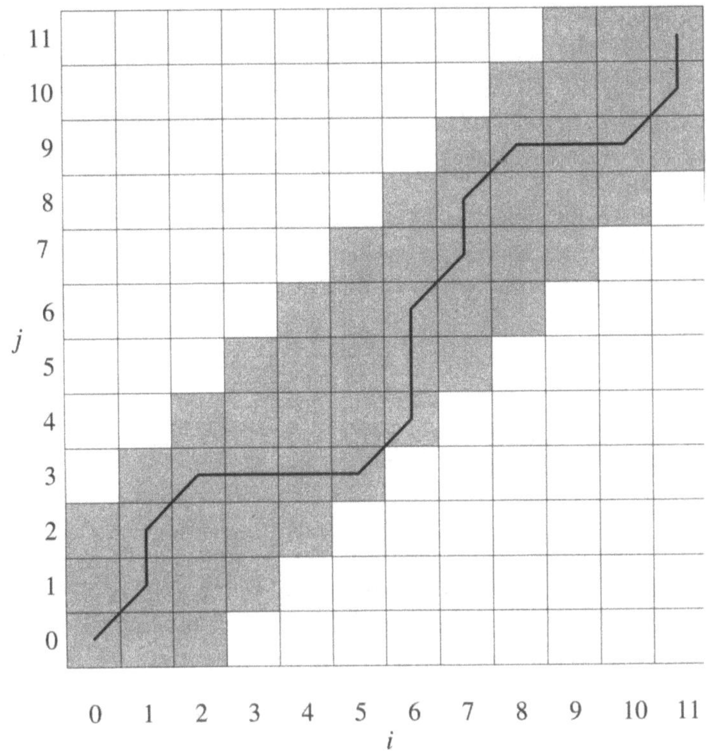

Figure 4.5. An example of a warping path with local constraint.

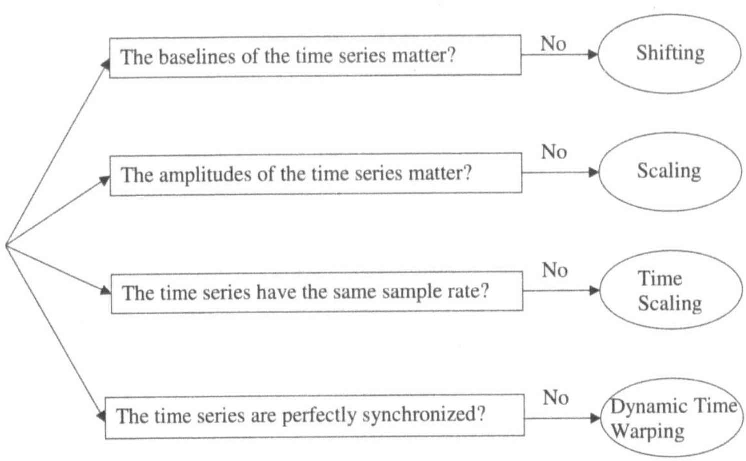

Figure 4.6. A decision tree for scaling, stretching, and warping transformations on time series.

4.5 Questions

1. Suppose that the answer to a time series similarity query is the time series $a, b, d, g, k, h, o, p, x$, and a program returns the approximate answer $a, b, c, g, h, i, j, o, p, x$. Which time series in the answer are false positives? Which are false negatives?
2. What are the definitions of recall and precision? What are the recall and precision for the approximate answers in question 1?
3. If we define the normal form $\tilde{\mathbf{x}}$ of a time series $\mathbf{x}$ as

$$\tilde{\mathbf{x}} = \frac{\mathbf{x} - \text{avg}(\mathbf{x})}{\sum(\mathbf{x} - \text{avg}(\mathbf{x}))^2},$$

   what is the relationship between corr($\mathbf{x}, \mathbf{y}$) and $D^2(\tilde{\mathbf{x}}, \tilde{\mathbf{y}})$?
4. Let $\mathbf{X}^n = DFT(\mathbf{x}^n)$ and $\mathbf{y}^{nw} = U^w(\mathbf{x}^n)$, that is

$$y_i = x_{\lfloor i/w \rfloor}, i = 0, 1, \ldots, nw - 1.$$

   What are the first $t, t \leq n$ DFT coefficients of $\mathbf{y}^{nw}$ in terms of $\mathbf{X}^n$?
5. Write a program to compute the LDTW(k) distance between time series.
   a) Compute the LDTW(2) distance between time series

$$0, 5, 8, 11, 11, 8, 5, 0, -4, -7, -8, -7, -5, -2, 1, 3, 3, 2, 0, -3$$

   and

$$0, 5, 8, 10, 9, 6, 2, -1, -1, 0, 3, 7, 9, 10, 7, 3, -2, -7.$$

   b) Compute their LDTW(5) distance.

Part II

Case Studies

5

StatStream

Summary. Consider the problem of monitoring tens of thousands of time series data streams in an online fashion and making decisions based on them. In addition to single stream statistics such as average and standard deviation, we also want to find the most highly correlated pairs of streams especially in a sliding window sense. A stock market trader might use such a tool to spot arbitrage opportunities. In this chapter, we propose efficient methods for solving this problem based on Discrete Fourier Transforms (see Chapter 2) and a three level time interval hierarchy. Extensive experiments on synthetic data and real world financial trading data show that our algorithm beats the direct computation approach by several orders of magnitude. It also improves on previous Fourier Transform approaches by allowing the efficient computation of time-delayed correlation over any size sliding window and any time delay. Correlation also lends itself to an efficient grid-based data structure. The result is the first algorithm that we know of to compute correlations over thousands of data streams in real time. The algorithm is incremental, has fixed response time, and can monitor the pairwise correlations of 10,000 streams on a single PC. The algorithm is embarrassingly parallelizable.

5.1 Introduction

Many applications consist of multiple data streams. For example,

- In mission operations for NASA's Space Shuttle, approximately 20,000 sensors are telemetered once per second to Mission Control at Johnson Space Center, Houston [59].
- There are about 50,000 securities trading in the United States, and every second up to 100,000 quotes and trades (ticks) are generated.

Unfortunately it is difficult to process such data in set-oriented data management systems, although object-relational time series extensions have begun to fill the gap [73]. For the performance to be sufficiently good however, "Data Stream Management Systems" (DSMSs) [13], whatever their logical model, should exploit the following characteristics of the application:

- Updates are through insertions of new elements (with relatively rare corrections of older data).

- Queries (moving averages, standard deviations, and correlation) treat the data as sequences not sets.
- Since a full stream is never materialized, queries treat the data as a never-ending data stream.
- One pass algorithms are desirable because the data is vast.
- Interpretation is mostly qualitative, so sacrificing accuracy for speed is acceptable.

Here we present the algorithms and architecture of StatStream, a data stream monitoring system. The system computes a variety of single and multiple stream statistics in one pass with constant time (per input) and bounded memory. To show its use for one practical application, we include most of the statistics that a securities trader might be interested in. The algorithms, however, are applicable to other disciplines, such as sensor data processing and medicine. The algorithms themselves are a synthesis of existing techniques and a few ideas of our own. We divide our contributions into functional and algorithmic. Our functional contributions are:

1. We compute multi-stream statistics such as synchronous as well as time-delayed correlation and vector inner product in a continuous online fashion. This means that if a statistic holds at time t, that statistic will be reported at time $t + v$, where v is a constant independent of the size and duration of the stream.
2. For any pair of streams, each pair-wise statistic is computed in an incremental fashion and requires constant time per update. This is done using a Discrete Fourier Transform approximation.
3. The approximation has a small error under natural assumptions.
4. Even when we monitor the data streams over sliding windows, no revisiting of the expiring data streams is needed.
5. The net result is that on a Pentium 4 PC, we can handle 10,000 streams each producing data every second with a delay window v of only 2 minutes.

Our algorithmic contributions mainly have to do with correlation statistics. First, we distinguish three time periods:

- Timepoint – the smallest unit of time over which the system collects data, e.g., second.
- Basic window – a consecutive subsequence of timepoints over which the system maintains a digest incrementally, e.g., a few minutes.
- Sliding window – a user-defined consecutive subsequence of basic windows over which the user wants statistics, e.g., an hour. The user might ask, "which pairs of stocks were correlated with a value of over 0.9 for the last hour?"

The use of the intermediate time interval that we call basic window yields three advantages:

1. Results of user queries need not be delayed more than the basic window time. In our example, the user will be told about correlations between 2 PM and 3 PM by 3:02 PM and correlations between 2:02 PM and 3:02 PM by 3:04 PM.

2. Maintaining stream digests based on the basic window allows the computation of correlations over windows of arbitrary size, not necessarily a multiple of basic window size, as well as time-delayed correlations with high accuracy.
3. A short basic window give fast response time but fewer time series can then be handled.

A second algorithmic contribution is the grid structure, each of whose cells stores the hash function value of a stream. The structure itself is unoriginal but the high efficiency we obtain from it is due to the fact that we are measuring correlation and have done the time decomposition mentioned above.

The remainder of this chapter will be organized as follows. The data we consider and statistics we produce are presented in Section 5.2. Section 5.3 presents our algorithms for monitoring high speed time series data streams. Section 5.4 discusses the system StatStream. Section 5.5 presents our experimental results. Section 5.6 puts our work in the context of related work.

5.2 Data And Queries

5.2.1 Time Series Data Streams

We consider data entering as a time ordered series of triples (streamID, timepoint, value). Each stream consists of all those triples having the same streamID. (In finance, a streamID may be a stock, for example.) The streams are synchronized.

Each stream has a new value available at every periodic time interval, e.g., every second. We call the interval index the **timepoint**. For example, if the periodic time interval is a second and the current timepoint for all the streams is i, after one second, all the streams will have a new value with timepoint $i+1$. (Note that if a stream has no value at a timepoint, a value will be assigned to that timepoint based on interpolation. If there are several values during a timepoint, then a summary value will be assigned to that timepoint.)

Let s_i or $s[i]$ denote the value of stream **s** at timepoint i. $s[i..j]$ denotes the subsequence of stream **s** from timepoints i through j inclusive. s^i denotes a stream with streamID i. Also we use t to denote the latest timepoint, i.e., now. The statistics we will monitor will be denoted $\text{stat}(s_j^{i_1}, s_j^{i_2}, \ldots, s_j^{i_k}, j \in [p, q])$, where the interval $[p, q]$ is a window of interest. We will discuss the meaning of windows in the next section.

5.2.2 Temporal Spans

In the spirit of the work in [35, 36], we generalize the three kinds of temporal spans for which the statistics of time series are calculated.

1. **Landmark windows:** In this temporal span, statistics are computed based on the values between a specific timepoint called landmark and the present. $\text{stat}(\mathbf{s}, \text{landmark}(k))$ will be computed on the subsequence of time series $s[i]$,

$i \geq k$. An unrestricted window is a special case when $k = 1$. For an unrestricted window the statistics are based on all the available data.
2. **Sliding windows:** In financial applications, at least, a sliding window model is more appropriate for data streams. Given the length of the sliding window w and the current timepoint t, stat(s, sliding(w)) will be computed in the subsequence $s[t - w + 1..t]$.
3. **Damped window model:** In this model, recent sliding windows are more important than previous ones. For example, in the computation of a moving average, a sliding window model will compute the average as $\text{avg} = \frac{\sum_{i=t-w+1}^{t} s_i}{w}$. By contrast, in a damped window model the weights of data decrease exponentially into the past. For example, a moving average in a damped window model can be computed as follows:

$$\text{avg}_{\text{new}} = \text{avg}_{\text{old}} * p + s_t * (1 - p), 0 < p < 1. \tag{5.1}$$

Other statistics in a damped window model can be defined similarly.

Here we will focus on the sliding window model, because it is the one used most often and is the most general.

5.2.3 Statistics to Monitor

Consider the stream $s_i, i = 1, \ldots, w$. The statistics we will monitor are as follows:

1. Single stream statistics, such as average, standard deviation, best fit slope. These are straightforward.
2. Correlation coefficients
3. Autocorrelation: the correlation of the series with itself at an earlier time.
4. Beta: the sensitivity of the values of a stream **s** to the values of another stream **r** (or weighted collection of streams). For example, in financial applications the beta measures the risk of a stock. A stock with a beta of 1.5 to the market index experiences 50 percent more movement in price than the market.

$$\text{beta}(s, r) = \frac{\frac{1}{w}\sum_{i=1}^{w} s_i r_i - \bar{s}\,\bar{r}}{\sum_{i=1}^{w}(r_i - \bar{r})^2}. \tag{5.2}$$

5.3 Statistics Over Sliding Windows

To compute the statistics over a sliding window, we will maintain a synopsis data structure for the stream to compute the statistics rapidly. To start, our framework subdivides the sliding windows equally into smaller windows, which we call **basic windows,** in order to facilitate the efficient elimination of old data and the incorporation of new data. We keep digests for both basic windows and sliding windows. For example, the running sum of the time series values within a basic window and the running sum within an entire sliding window belong to the two kinds of digests

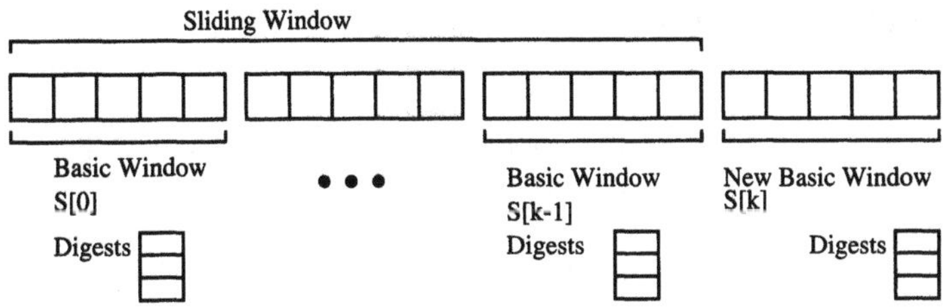

Figure 5.1. Sliding windows and basic windows.

respectively. Figure 5.1 shows the relation between sliding windows and basic windows.

Let the data within a sliding window be $s[t - w + 1..t]$. Suppose $w = kb$, where b is the length of a basic window and k is the number of basic windows within a sliding window. Let $S[0], S[1], \ldots, S[k-1]$ denote a sequence of basic windows, where $S[i] = s[(t - w) + ib + 1..(t - w) + (i + 1)b]$. $S[k]$ will be the new basic window and $S[0]$ is the expiring basic window. The j-th value in the basic window $S[i]$ is $S[i; j]$. Table 5.1 defines the symbols that are used in the rest of the chapter.

Table 5.1. Symbols.

w	Size of a sliding window
b	Size of a basic window
k	Number of basic windows within a sliding window
n	Number of DFT coefficients used as digests
N_s	Number of data streams

The size of the basic window is important because it must be possible to report all statistics for basic window i to the user before basic window $i + 1$ completes (at which point it will be necessary to begin computing the statistics for window $i + 1$).

5.3.1 Single Stream Statistics

Single stream statistics such as moving average and moving standard deviation are computationally inexpensive. In this section, we discuss moving averages to demonstrate the concept of maintaining digest information based on basic windows. Obviously, the information to be maintained for the moving average is $\sum(s[t-w+1..t])$. For each basic window $S[i]$, we maintain the digest $\sum(S[i]) = \sum_{j=1}^{b} S[i; j]$. After b new data points from the stream become available, we compute the sum over the new basic window $S[k]$. The sum over the sliding window is updated as follows:

$$\sum_{\text{new}}(s) = \sum_{\text{old}}(s) + \sum S[k] - \sum S[0].$$

5.3.2 Correlation Statistics

Correlation statistics are important in many applications. For example, Pairs Trading, also known as correlation trading strategy, is widely employed by major Wall Street firms. This strategy focuses on trading pairs of equities that are correlated. The correlation between two streams (stocks) is affected by some factors that are not known a priori. Any pair of streams could be correlated at some time. Much effort has been made to find such correlations in order to enjoy arbitrage profits. These applications imply that the ability to spot correlations among a large number of streams in real time will provide competitive advantages. To make our task even more challenging, such correlations change over time and we have to update the moving correlations frequently.

The efficient identification of highly correlated streams potentially requires the computation of all pairwise correlations and could be proportional to the total number of timepoints in each sliding window times all pairs of streams. We make this computation more efficient by (1) using a discrete fourier transform of basic windows to compute the correlation of stream pairs approximately; (2) using a grid data structure to avoid the approximate computation for most pairs.

We will explain how to compute the vector inner product when the two series are aligned in basic windows. This approach is extended to the general case when the basic windows are not aligned. Then we will show our approach for reporting the highly correlated stream pairs in an online fashion.

5.3.3 Inner Product with Aligned Windows

The correlation and beta can be computed from the vector inner product. Remember that the vector inner product for two series $\mathbf{x} = (x_1, \ldots, x_w), \mathbf{y} = (y_1, \ldots, y_w)$, denoted as $\psi(\mathbf{x}, \mathbf{y})$, is just the sum of the products of their corresponding elements:

$$\psi(\mathbf{x}, \mathbf{y}) = \sum_{i=1}^{w} x_i y_i^*. \tag{5.3}$$

Given two series $\mathbf{s}^x$ and $\mathbf{s}^y$, when the two series are aligned,

$$\psi(\mathbf{s}^x, \mathbf{s}^y) = \sum_{i=1}^{k} \psi(S^x[i], S^y[i]). \tag{5.4}$$

This is illustrated in Figure 5.2. So, we must explain how to compute the inner product of two basic windows: $x_1, x_2, \ldots, x_b$ and $y_1, y_2, \ldots, y_b$.

Let $f : f_1(\mathbf{x}), f_2(\mathbf{x}), \ldots$ be a family of continuous functions. We approximate the time series in each basic window, $S^x[i] = x_1, x_2, \ldots, x_b$ and $S^y[i] = y_1, y_2, \ldots, y_b$, with a function family f in the spirit of Chapter 2. (We will give specific examples later.)

$$x_i \approx \sum_{m=0}^{n-1} c_m^x f_m(i), \quad y_i \approx \sum_{m=0}^{n-1} c_m^y f_m(i) \qquad i = 1, \ldots, b, \tag{5.5}$$

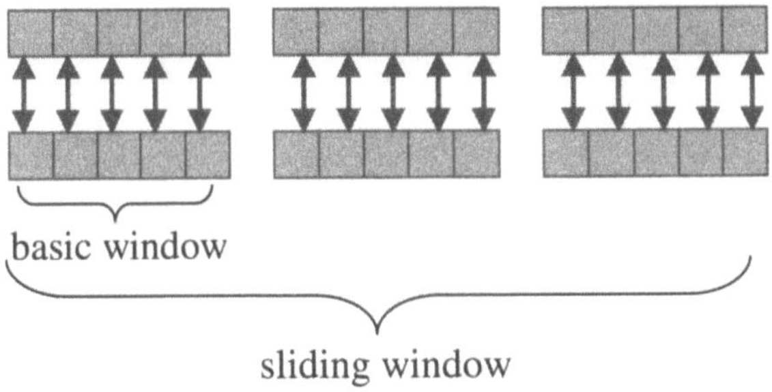

Figure 5.2. Illustrated computation of inner product with aligned windows.

where $c_m^x, c_m^y, m = 0, \ldots, n-1$ are n coefficients to approximate the time series with the function family f.

The inner product of the two basic windows is therefore

$$\begin{aligned}\sum_{i=1}^{b} x_i y_i^* &\approx \sum_{i=1}^{b} \Big(\sum_{m=0}^{n-1} c_m^x f_m(i) \sum_{p=0}^{n-1} c_p^{y*} f_p^*(i) \Big) \\ &= \sum_{m=0}^{n-1} \sum_{p=0}^{n-1} c_m^x c_p^{y*} \Big(\sum_{i=1}^{b} f_m(i) f_p^*(i) \Big) \\ &= \sum_{m=0}^{n-1} \sum_{p=0}^{n-1} c_m^x c_p^{y*} W(m, p). \end{aligned} \tag{5.6}$$

In (5.6), $W(m, p) = \sum_{i=1}^{b} f_m(i) f_p^*(i)$ can be precomputed. If the function family f is orthogonal, we have

$$W(m, p) = \begin{cases} 0, & m \neq p \\ V(m) \neq 0, & m = p. \end{cases} \tag{5.7}$$

Thus,

$$\sum_{i=1}^{b} x_i y_i^* \approx \sum_{m=0}^{n-1} c_m^x c_m^{y*} V(m). \tag{5.8}$$

With this curve fitting technique, we reduce the space and time required to compute inner products from b to n. It should also be noted that besides data compression, the curve fitting approach can be used to fill in missing data. This works naturally for computing correlations among streams with missing data at some timepoints.

The sine-cosine function families have the right properties. We perform the Discrete Fourier Transforms for the time series over the basic windows, enabling a constant time computation of coefficients for each basic window. Following the observations in [8], we can obtain a good approximation for the series with only the first n DFT coefficients

$$x_i \approx \frac{1}{\sqrt{b}} \sum_{F=0}^{n-1} X_F e^{j2\pi Fi/b} \qquad i = 1, 2, \ldots, b. \tag{5.9}$$

Let

$$f_m(i) = \frac{1}{\sqrt{b}} e^{j2\pi mi/b};$$

the users can verify that

$$\sum_{i=1}^{b} f_m(i) f_p^*(i) = \begin{cases} 0, & m \neq p \\ V(m) \neq 0, & m = p. \end{cases}$$

5.3.4 Inner Product with Unaligned Windows

A much harder problem is to compute correlations with time lags. The time series will not necessarily be aligned at their basic windows. However, the digests we keep are enough to compute such correlations.

Without loss of generality, we will show the computation of the n-approximate lagged inner product of two streams with time lags less than the size of a basic window.

Given two such series,

$$\mathbf{s}^x = s_1^x, \ldots, s_w^x = S^x[1], \ldots, S^x[k]$$

and

$$\mathbf{s}^y = s_{a+1}^y, \ldots, s_{a+w}^y = S^y[1], \ldots, S^y[k], S^y[k+1],$$

where for the basic windows $S^y[1]$ and $S^y[k+1]$, only the last $b-a$ values in $S^y[2]$ and the first a values in $S^y[k+1]$ are included ($a < b$). We have

$$\begin{aligned} \psi(\mathbf{s}^x, \mathbf{s}^y) &= \sum_{i=1}^{w} (s_i^x s_{a+i}^y) \\ &= \sum_{j=1}^{k} (\rho(S^x[j], S^y[j], b-a) + \rho(S^y[j+1], S^x[j], a), \end{aligned} \tag{5.10}$$

where $\rho(S^1, S^2, d) = \sum_{i=1}^{d} S^1[i] S^2[b-d+i]$. Figure 5.3 illustrates the computation.

For $S^1 = x_1, \ldots, x_b$ and $S^2 = y_1, \ldots, y_b$, $\rho(S^1, S^2, d)$ is the inner product of the first d values of S^1 with the last d values of S^2.This can be approximated using only their first n DFT coefficients. Whereas in the aligned case, the product of terms pertaining to different frequencies is zero. That does not hold for the unaligned case.

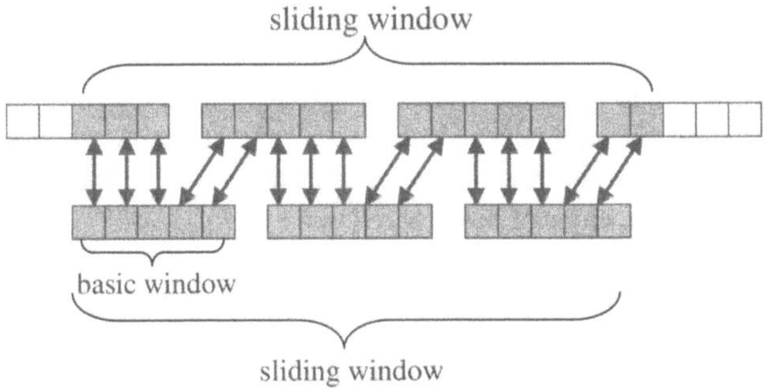

Figure 5.3. Illustrated computation of inner product with unaligned windows.

$$\begin{aligned}\sum_{i=1}^{d} x_i y_{b-d+i}^* &\approx \sum_{i=1}^{d}\Big(\sum_{m=0}^{n-1} c_m^x f_m(i) \sum_{p=0}^{n-1} c_p^{y*} f_p^*(b-d+i)\Big)\\ &= \sum_{m=0}^{n-1}\sum_{p=0}^{n-1} c_m^x c_p^{y*}\Big(\sum_{i=1}^{d} f_m(i) f_p^*(b-d+i)\Big)\\ &= \sum_{m=0}^{n-1}\sum_{p=0}^{n-1} c_m^x c_p^{y*} W(m,p,d). \end{aligned} \tag{5.11}$$

This implies that if we precompute the table of

$$W(m,p,d) = \sum_{i=1}^{d} f_m(i) f_p^*(b-d+i) \tag{5.12}$$

where $m, p = 0 \ldots, n-1; d = 1, \ldots, \lfloor b/2 \rfloor$, we can compute the inner product using only the DFT coefficients without requiring the alignment of basic windows in time $O(n^2)$ for each basic window. Precomputation time for any specific displacement d is $O(n^2 d)$.

Theorem 5.1. *The n-approximate lagged inner product of two time series using their first n DFT coefficients can be computed in time $O(kn)$ if the two series have aligned basic windows, otherwise it takes time $O(kn^2)$, where k is the number of basic windows.*

It is not hard to show that this approach can be extended to compute the inner product of two time series over sliding windows of any size.

Corollary 5.2. *The inner product of two time series over sliding windows of any size with any time delay can be approximated using only the basic window digests of the data streams.*

5.3.5 IO Performance

It might be desirable to store the summary data for future analysis. Since the summary data we keep are sufficient to compute all the statistics we are interested in, there is no need to store the raw data streams. The summary data will be stored on disk sequentially in the order of basic windows.

Let N_s be the number of streams and let n be the number of DFT coefficients we use ($2n$ real number), the I/O cost to access the data for all streams within a specific period will be

$$\frac{N_s\ k\ \text{Size of(float)}(2+2n)}{\text{Page size}}$$

while the I/O cost for the exact computation is

$$\frac{N_s\ k\ \text{Size of(float)}b}{\text{Page size}}.$$

The improvement is a ratio of $\frac{b}{2+2n}$. The first 2 corresponds to two non-DFT elements of summary data: the sum and the sum of the squares of the time series in each basic window. Also the I/O costs above assume sequential disk access. This is a reasonable assumption given the time-ordered nature of data streams.

5.3.6 Monitoring Correlations between Data Streams

The above curve fitting technique can be used for computing inner products and correlations over sliding windows of any size, as well as with any time delay across streams. A frequent goal is to discover streams with high correlations. To do online monitoring of synchronized streams over a fixed size of sliding window with correlation above a specific threshold, we use an approach based on DFT and a hash technique that will report such stream pairs quickly.

From Lemma 4.7, we can reduce the correlation coefficient to Euclidean Distance, and therefore we can apply the techniques in [8] to report sequences with correlation coefficients higher than a specific threshold.

Lemma 5.3. *Let the normalized form of two time series* $\mathbf{x}$ *and* $\mathbf{y}$ *be* $\hat{\mathbf{y}}$ *and* $\hat{\mathbf{x}}$ *respectively, also the Discrete Fourier Transform of* $\hat{\mathbf{x}}$ *and* $\hat{\mathbf{y}}$ *are* $\hat{\mathbf{X}}$ *and* $\hat{\mathbf{Y}}$ *respectively,*

$$\text{corr}(\mathbf{x},\mathbf{y}) \geq 1-\epsilon^2 \Rightarrow D_n(\hat{\mathbf{X}},\hat{\mathbf{Y}}) \leq \sqrt{w}\epsilon, \tag{5.13}$$

where $D_n(\hat{\mathbf{X}},\hat{\mathbf{Y}})$ *is the Euclidean distance between series* $\hat{X}_1,\hat{X}_2,\ldots,\hat{X}_n$ *and* $\hat{Y}_1,\hat{Y}_2,\ldots,\hat{Y}_n$.

Proof. As the Discrete Fourier Transform preserves Euclidean distance (Theorem 2.24), we have

$$D(\hat{\mathbf{X}},\hat{\mathbf{Y}}) = D(\hat{\mathbf{x}},\hat{\mathbf{y}}).$$

Using only the first n and last n, ($n \ll w$) DFT coefficients [8, 82], from the symmetry property of DFT, we have

$$\sum_{i=1}^{n}(\hat{X}_i - \hat{Y}_i)^2 + \sum_{i=1}^{n}(\hat{X}_{w-n} - \hat{Y}_{w-n})^2 = 2\sum_{i=1}^{n}(\hat{X}_i - \hat{Y}_i)^2$$
$$= 2D_n^2(\hat{\mathbf{X}}, \hat{\mathbf{Y}})$$
$$\leq D^2(\hat{\mathbf{X}}, \hat{\mathbf{Y}}).$$

From Lemma 4.7, we know that

$$D^2(\hat{\mathbf{x}}, \hat{\mathbf{y}}) = 2w\big(1 - \text{corr}(\mathbf{x}, \mathbf{y})\big).$$

Therefore,

$$\begin{aligned} & \text{corr}(x, y) \geq 1 - \epsilon^2 \\ \Rightarrow\quad & D^2(\hat{\mathbf{x}}, \hat{\mathbf{y}}) \leq 2w\epsilon^2 \\ \Rightarrow\quad & D^2(\hat{\mathbf{X}}, \hat{\mathbf{Y}}) \leq 2w\epsilon^2 \\ \Rightarrow\quad & D_n(\hat{\mathbf{X}}, \hat{\mathbf{Y}}) \leq \sqrt{w}\epsilon. \end{aligned}$$

□

From the above lemma, we can examine the correlations of only those stream pairs for which $D_n(\hat{\mathbf{X}}, \hat{\mathbf{Y}}) \leq \sqrt{w}\epsilon$ holds. We will get a superset of highly correlated pairs and there will be no false negatives. The false positives can be further filtered out as we explain later. HierarchyScan [66] also uses correlation coefficients as a similarity measure for time series. They use $\mathcal{F}^{-1}\{\hat{X}_i^* \hat{Y}_i\}$ as an approximation for the correlation coefficient. Because such approximations will not be always above the true values, some stream pairs could be discarded based on their approximations, even if their true correlations are above the threshold. Though HierarchyScan propose an empirical method to select level-dependent thresholds for multi-resolution scans of sequences, it cannot guarantee the absence of false negatives.

We can extend the techniques above to report stream pairs of negative high-value correlations.

Lemma 5.4. *Let the Discrete Fourier Transform of the normalized form of two time series $\hat{\mathbf{x}}$ and $\hat{\mathbf{y}}$ be $\hat{\mathbf{X}}$ and $\hat{\mathbf{Y}}$.*

$$\text{corr}(\mathbf{x}, \mathbf{y}) \leq -1 + \epsilon^2 \Rightarrow D_n(-\hat{\mathbf{X}}, \hat{\mathbf{Y}}) \leq \sqrt{w}\epsilon. \tag{5.14}$$

Proof. We have

$$DFT(-\mathbf{x}) = -DFT(\mathbf{x})$$

$$\begin{aligned} & \text{corr}(\mathbf{x}, \mathbf{y}) \leq -1 + \epsilon^2 \\ \Rightarrow\quad & \sum_{i=1}^{w} \hat{x}_i \hat{y}_i \leq -1 + \epsilon^2 \\ \Rightarrow\quad & \sum_{i=1}^{w} (-\hat{x}_i)\hat{y}_i \geq 1 - \epsilon^2 \\ \Rightarrow\quad & D_n(-\hat{\mathbf{X}}, \hat{\mathbf{Y}}) \leq \sqrt{w}\epsilon. \end{aligned}$$

□

Now we discuss how to compute the DFT coefficients $\hat{\mathbf{X}}$ incrementally. The DFT coefficients $\hat{\mathbf{X}}$ of the normalized sequence can be computed from the DFT coefficients $\mathbf{X}$ of the original sequence.

Lemma 5.5. *Let* $\hat{\mathbf{X}} = DFT(\hat{\mathbf{x}})$, $\hat{\mathbf{Y}} = DFT(\hat{\mathbf{y}})$*; we have*

$$\begin{cases} \hat{X}_0 = 0 \\ \hat{X}_i = \frac{X_i}{\mathrm{std}(\mathbf{x})} \quad i \neq 0. \end{cases} \tag{5.15}$$

We can maintain DFT coefficients over sliding windows incrementally [39].

Lemma 5.6. *Let* X_m^{old} *be the m-th DFT coefficient of the series in sliding window* $x_0, x_1, \ldots, x_{w-1}$ *and* X_m^{new} *be that coefficient of the series* $x_1, x_2, \ldots, x_w$,

$$X_m^{\mathrm{new}} = e^{j2\pi m/w}\left(X_m^{\mathrm{old}} + \frac{x_w - x_0}{\sqrt{w}}\right) \qquad m = 1, \ldots, n. \tag{5.16}$$

This can be extended to a batch update based on the basic windows.

Lemma 5.7. *Let* X_m^{old} *be the m-th DFT coefficient of the series in sliding window* $x_0, x_2, \ldots, x_{w-1}$ *and let* X_m^{new} *that coefficient of the series* $x_b, x_{b+1}, \ldots, x_w, x_{w+1}, \ldots, x_{w+b-1}$,

$$X_m^{\mathrm{new}} = e^{j2\pi mb/w} X_m^{\mathrm{old}} + \frac{1}{\sqrt{w}}\left(\sum_{i=0}^{b-1} e^{j2\pi m(b-i)/w} x_{w+i} - \sum_{i=0}^{b-1} e^{j2\pi m(b-i)/w} x_i\right) \tag{5.17}$$

where $m = 1, \ldots, n$.

The corollary suggests that to update the DFT coefficients incrementally, we should keep the following n digests for the basic windows:

$$\zeta_m = \sum_{i=0}^{b-1} e^{j2\pi m(b-i)/w} x_i, \qquad m = 1, \ldots, n. \tag{5.18}$$

These digests are the components of the DFT of the sliding window in the basic window.

By using the DFT on normalized sequences, we also map the original sequences into a bounded feature space.

Lemma 5.8. *Let* $\hat{X}_0, \hat{X}_1, \ldots, \hat{X}_{w-1}$ *be the Discrete Fourier Transform of a normalized sequence* $x_1, x_2, \ldots, x_w$*; we have*

$$|\hat{X}_i| \leq \frac{\sqrt{2w}}{2}, \qquad i = 1, \ldots, n, n < w/2. \tag{5.19}$$

Proof.

$$\sum_{i=1}^{w-1} (\hat{X}_i)^2 = \sum_{i=1}^{w} (\hat{x}_i)^2 = w$$

$$\Rightarrow \quad 2\sum_{i=1}^{n} \hat{X}_i^2 = \sum_{i=1}^{n} (\hat{X}_i^2 + \hat{X}_{w-i}^2) \le w$$

$$\Rightarrow \quad |\hat{X}_i| \le \frac{\sqrt{2w}}{2}, \qquad i = 1, \ldots, n. \qquad \square$$

From the above lemma, each DFT coefficient ranges from $-\frac{\sqrt{2w}}{2}$ to $\frac{\sqrt{2w}}{2}$, therefore the DFT feature space R^{2n} is a cube of diameter $\sqrt{2w}$. We can use a grid structure [16] to report near neighbors efficiently.

We will use the first $\hat{n}$, $\hat{n} \le 2n$, dimensions of the DFT feature space for indexing. We superimpose an $\hat{n}$-dimensional orthogonal regular grid on the DFT feature space and partition the cube of diameter $\sqrt{2w}$ into cells with the same size and shape. There are $(2\lceil \frac{\sqrt{2}}{2\epsilon} \rceil)^{\hat{n}}$ cells of cubes of diameter $\sqrt{w}\epsilon$. Each stream is mapped into a cell based on its first $\hat{n}$ normalized DFT coefficients. Suppose that a stream $\mathbf{x}$ is hashed to cell $(c_1, c_2, \ldots, c_{\hat{n}})$. To report the streams whose correlation coefficients with $\mathbf{x}$ are above the threshold $1 - \epsilon^2$, only streams hashed to cells adjacent to cell $(c_1, c_2, \ldots, c_{\hat{n}})$ need to be examined. Similarly, streams whose correlation coefficients with $\mathbf{x}$ are less than the threshold $-1 + \epsilon^2$, must be hashed to cells adjacent to cell $(-c_1, -c_2, \ldots, -c_{\hat{n}})$. After hashing the streams to cells, the number of stream pairs to be examined is greatly reduced. We can then compute their Euclidean distance, as well as correlation, based on the first n DFT coefficients.

For example, given many high-speed time series streams that are updated every second, we want to monitor their correlation over a two hour sliding window. To do that, we will maintain the first $n = 16$ normalized DFT coefficients of each time series stream over the sliding window. Suppose that we choose a basic window size of five minutes. There will be 300 timepoints in a basic window and 24 basic windows in a sliding window. For each basic window of a time series, we need to maintain the 16 digests in (5.18), $\zeta_1, \zeta_2, \ldots, \zeta_{16}$. From these digests in the basic windows, we can maintain the first 16 normalized DFT coefficients of a time series over a sliding window incrementally. Out of the 16 DFT coefficients, only the first $\hat{n} = 4$ coefficients are used for indexing in the grid structure. If the grid report that a pair of streams is likely to the correlated, we then check their correlation based on their 16 DFT coefficients.

The grid structure can be maintained as follows. Without loss of generality, we discuss the detection of only positive high-value correlations. There are two kinds of correlations the user might be interested in.

- **Synchronized correlation** If we are interested in only synchronized correlation, the grid structure is cleared at the end of every basic window. At the end of each basic window, all the data within the current basic window are available and the digests are computed. Suppose that stream x is hashed to cell c; then x will be compared to any stream that has been hashed to the neighborhood of c.

- **Lagged correlation** If we also want to report lagged correlation, including autocorrelation, the maintenance of the grid structure will be a little more complicated. Let T_M be a user-defined parameter specifying the largest lag that is of interest. Each cell in the grid as well as the streams hashed to the grid will have a timestamp. The timestamp T_x associated with the hash value of the stream x is the time when x is hashed to the grid. The timestamp T_c of a cell c is the latest timestamp when the cell c is updated. The grid is updated every basic window time but never globally cleared. Let S_c be the set of cells that are adjacent to c, including c. Figure 5.4 gives the pseudo-code to update the grid when stream x hashes to cell c.

```
FOR ALL cells c_i ∈ S_c
    IF T_x − T_{c_i} > T_M  //T_{c_i} is the timestamp of c_i.
        clear(c_i) //all the streams in c_i are out of date.
    ELSE IF 0 < T_x − T_{c_i} ≤ T_M
    //some streams in c_i are out of date.
        FOR ALL stream y ∈ c_i
            IF T_x − T_y > T_M delete(y)
            ELSE examine_correlation(x, y)
    ELSE //T_x = T_{c_i}
        FOR ALL stream y ∈ c_i
            examine_correlation(x, y)
    T_{c_i} = T_x //to indicate that c_i is just updated
```

Figure 5.4. Algorithm to detect lagged correlation.

Theorem 5.9. *Given a collection of time series streams, using only the digests of the streams, our algorithms can find those stream pairs whose correlations (whether synchronized or lagged) are above a threshold without false negatives.*

Using the techniques in this section, we can search for high-value lagged correlations among data streams very fast. The time lag must be a multiple of the size of a basic window in this case. Theorem 5.1 states that we can approximate correlation of any time lag efficiently. Approximation methods are used as a post-processing step after the hashing methods spot those promising stream pairs.

5.3.7 Parallel Implementation

Our framework facilitates a parallel implementation by using a straightforward decomposition. Consider a network of K servers to monitor N_s streams. We assume these servers have similar computing resources.

The work to monitor the streams consists of two stages.

1. Compute the digests and single stream statistics for the data streams. The N_s streams are equally divided into K groups. The server $i\,(i = 1, \ldots, K)$ will read those streams in the i-th group and compute their digests, single stream statistics and hash values.
2. Report highly correlated stream pairs based on the grid structure. The grid structure is also geometrically and evenly partitioned into K parts. A server X will read in its part, a set of cells S_X. Server X will also read a set of cells S'_X including cells adjacent to the boundary cells in S_X. Server X will report those stream pairs that are highly correlated within cells in S_X. Note that only the first n normalized DFT coefficients need to be communicated between servers, thus reducing the overhead for communication.

5.4 Software Architecture of the StatStream System

StatStream runs in a high performance interpreted environment called K [1]. Our system makes use of this language's powerful array-based computation to achieve high speed in the streaming data environment. Because K is supported in different platforms such as Windows, Linux and Solaris, our system based on K runs on these platforms too. The system follows the algorithmic ideas above and makes use of the following parameters:

- **Correlation Threshold** Only stream pairs whose correlation coefficient is larger than a specified threshold will be reported. The higher this threshold, the finer the grid structure, and the fewer streams whose exact correlations must be computed.
- **Sliding Window Size** This is the time interval over which statistics are reported to the user. If the sliding window size is 1 hour, then the reported correlations are those over the past hour.
- **Duration over Threshold** Some users might be interested in only those pairs with correlation coefficients above the threshold for a pre-defined period. For example, a trader might ask "Has the one hour correlation between two stocks been over 0.95 during the last 10 minutes?" This parameter provides such users with an option to specify a minimum duration. A longer duration period of highly correlated streams indicates a stronger relationship between the streams while a shorter one might indicate an accidental correlation. For example, a longer duration might give a stock market trader more confidence when taking advantage of such potential opportunities. A longer duration also gives better performance because we can update the correlations less frequently.
- **Range of Lagged Correlations** In addition to synchronized correlations, StatStream can also detect high-value lagged correlations. This parameter, i.e., T_M in Section 5.3.6, specifies the range of the lagged correlations. For example, if the range is 10 minutes and the basic window is 2 minutes, the system will examine cross-correlations and autocorrelations for streams with lags of 2,4,8 and 10 minutes.

Figure 5.5 shows the software architecture of StatStream. You can find the download and running instructions at the following website:

http://www.cs.nyu.edu/cs/faculty/shasha/papers/statstream.html

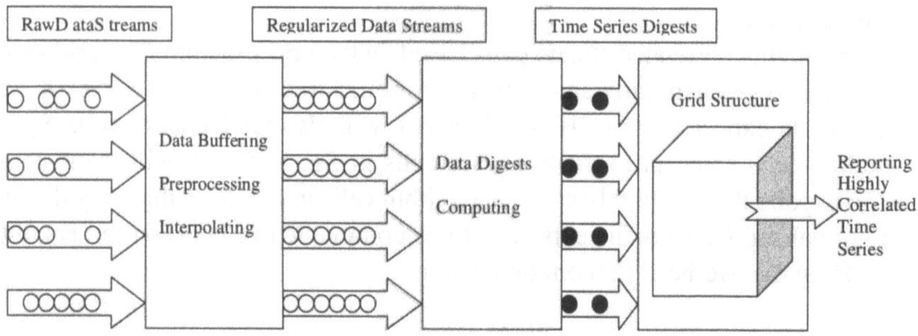

Figure 5.5. The software architecture of StatStream.

5.5 Empirical Study

Our empirical studies attempt to answer the following questions.

- How great are the time savings when using the DFT approximate algorithms as compared with exact algorithms? How many streams can they handle in real time?
- What's the approximation error when using DFT within each basic window to estimate correlation? How does it change according to the basic and sliding window sizes?
- What is the pruning power of the grid structure in detecting high correlated pairs? What is the precision?

We perform the empirical study on the following two datasets on a 1.5G Hz Pentium 4 PC with 128 MB of main memory.

- **Synthetic Data** Time series streams are generated using the random walk model. For stream s,

$$s_i = 100 + \sum_{j=1}^{i}(u_j - 0.5), \qquad i = 1, 2, \ldots$$

  where u_j is a set of uniform random real numbers in [0, 1].
- **Stock Exchange Data** The New York Stock Exchange (NYSE) Trade and Quote (TAQ) database provides intraday trade and quote data for all stocks listed on NYSE, AMEX, NASDAQ, and SmallCap issues. The database grows at the rate of 10GB per month. The historical data since 1993 have accumulated to 500GB. The data we use in our experiment are the tick data of the major stocks in a trading day. The 300 stocks in this dataset are heavily traded in NYSE. During the peak

hours, there several trades for each stock in a second. We use the price weighted by volume as the price of that stock at that second. In a second when there is no trading activities for a particular stock, we use the last trading price as its price. In this way, all the stock streams are updated every second, corresponding to a timepoint. The sliding window will vary from half an hour to two hours (1,800 to 7,200 timepoints). In practice the actual choice of the sliding windows will be up to the user. The lengths of the basic windows are half a minute to several minutes, depending on the number of streams to be monitored and the computing capacity.

5.5.1 Speed Measurement

Suppose that the streams have new data every second. The user of a time series stream system asks himself the following questions:

1. How many streams can I track at once in an online fashion? (Online means that even if the data come in forever, I can compute the statistics of the data with a fixed delay from their occurrence.)
2. How long is the delay between a change in correlation and the time when I see the change?

Our system will compute correlations at the end of each basic window. As noted above, the computation for basic window i must finish by the end of basic window $i+1$ in order for our system to be considered on-line. Otherwise, it would lag farther and farther behind over time. Therefore, some of the correlations may be computed towards the end of the basic window $i+1$. The user perceives the size of the basic window as the maximum delay between the time that a change in correlation takes place and the time it is computed.

The net result is that the answers to questions (1) and (2) are related. We can increase the number of streams at the cost of increasing the delay in reporting correlation.

Figure 5.6 shows the number of streams vs. the minimum size of the basic window for a uniprocessor and with different algorithms. In the DFT method, we choose the number of coefficients to be 16.

Using the exact method, given the basic window size b, the time to compute the correlations among N_s streams with b new timepoints is $T = k_0 b N_s^2$. Because the algorithm must finish this in b seconds, we have $k_0 b N_s^2 = b \Rightarrow N_s = \sqrt{\frac{1}{k_0}}$.

With the DFT-grid method, the work to monitor correlations has two parts: (1) Updating digests takes time $T_1 = k_1 b N_s$; (2) Detecting correlation based on the grid takes time $T_2 = k_2 N_s^2$. To finish these two computations before the basic window ends, we have $T_1 + T_2 = b$. Since T_2 is the dominating term, we have $N_s \approx \sqrt{\frac{b}{k_2}}$. Note that because of the grid structure, $k_2 \ll k_0$. Also, the computation with data digests is much more IO efficient than the exact method on the raw streams. From the equation above, we can increase the number of streams monitored by increasing the basic window size, i.e., delay time. This tradeoff between response

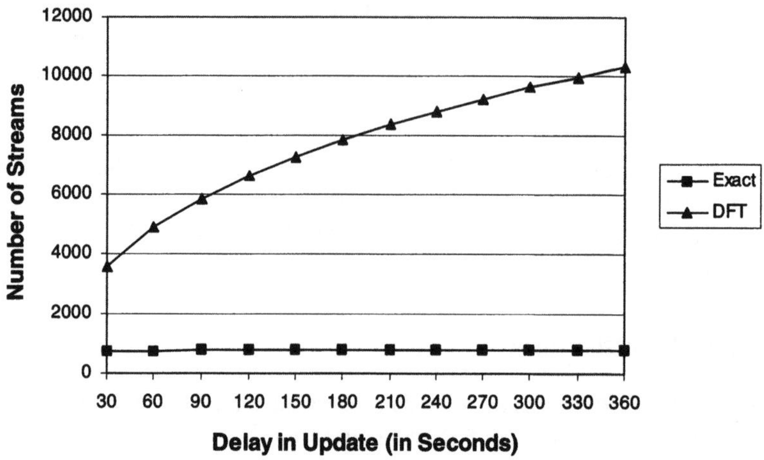

Figure 5.6. Comparison of the number of streams the DFT and exact methods can handle.

time and throughput is confirmed in the figure. The number of streams handled by our system increases with the size of the basic window, while there is no perceivable change for the exact algorithm.

The wall clock time to find correlations using the DFT-grid method is much faster than the exact method (Figure 5.7, *top*). The time is divided into two parts: detecting correlation and updating the digest (Figure 5.7, *bottom*).

The experiments on the processing time for the DFT-grid method also provide a guideline on how to choose the size of the basic window. Given a specific computing capacity, Figure 5.8 shows the processing time for different basic window sizes, when the numbers of streams are 5,000 (Figure 5.8a) and 10,000 (Figure 5.8b). Note that the wall clock time to detect correlation does not change with the size of the basic window, while the wall clock time to update the digest is proportional to the size of the basic window. Therefore the processing time is linear to the size of the basic window. Because the processing time must be shorter than the basic window time for our system to be online, we can decide the minimum basic window size. From Figure 5.8 to monitor 5,000 streams the minimum basic window size is 25 seconds. Similarly, the basic window time should be no less than 150 seconds for 10,000 streams.

5.5.2 Measuring Precision

Because the approximate correlations are based on DFT curve fitting in each basic window, the precision of the computation depends on the size of the basic window. Our experiments on the two data sets (Figure 5.9) show that the error (measured by the absolute value of the difference to the real correlation coefficients) increases with larger basic window size and decreases with larger sliding window size, but remain

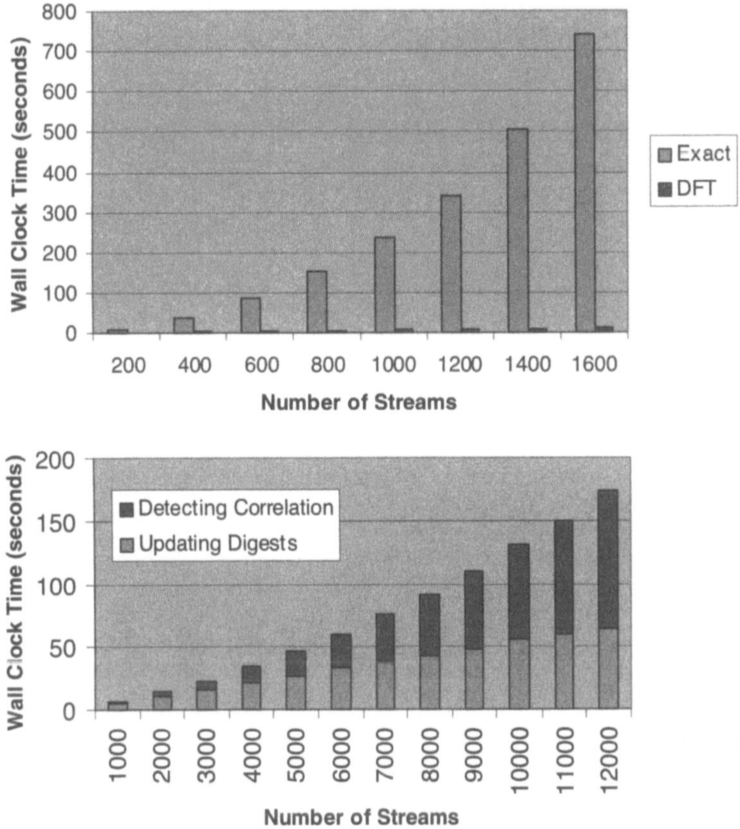

Figure 5.7. Wall clock time comparisons.

small throughout. This is particularly noteworthy, because we used only the first two DFT coefficients in each basic window.

We also performed experiments to test the effectiveness of the grid structure. The grid structure on DFT feature space prunes out most of the low-correlated pairs of streams. The pruning power [54] is the number of pairs reported by the grid, divided by the number of all potential pairs. Since our system guarantees no false negatives, the reported pairs include all the high-correlated pairs and some false positives. We also measure the quality of the system by precision, which is the ratio of the number of pairs whose correlations are above the threshold and reported by StatStream, to the number of pairs that are reported by StatStream. Figure 5.10 shows the precision and pruning power fraction using different numbers of coefficients and values of thresholds for different datasets. $R0.85$ indicates the real dataset with threshold of 0.85, $S0.9$ indicates the synthetic dataset with threshold of 0.9, etc. The length of the sliding window is one hour.

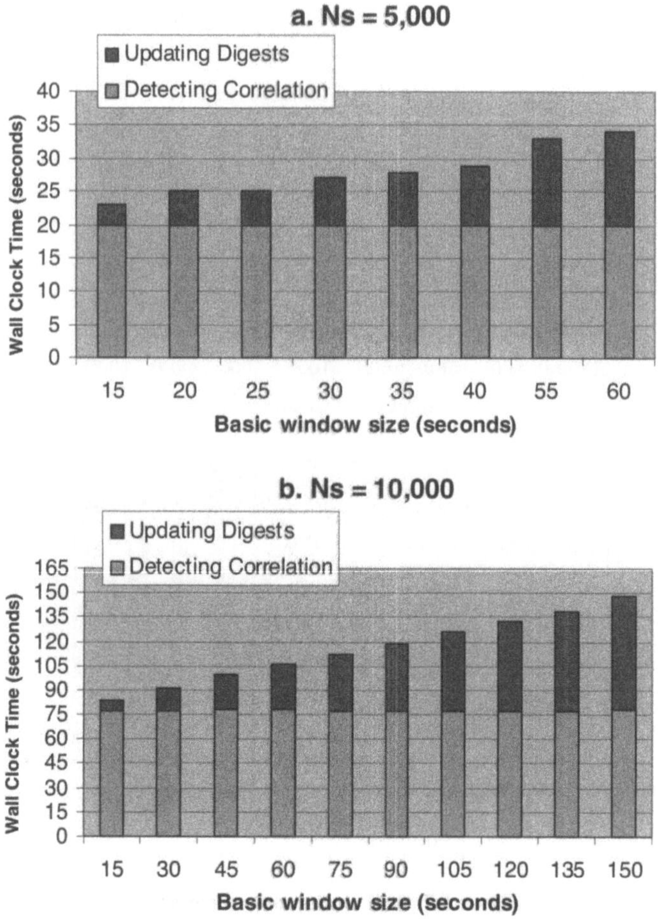

Figure 5.8. Wall clock time comparisons for different basic window sizes.

Table 5.2 shows that recall can be traded off against precision. The user may specify a tolerance t such that the system will report those pairs with approximate correlation coefficients larger than (threshold $-$ t).

Table 5.2. Precision after post processing.

Dataset	R0.85	R0.85	R0.9	R0.9	S0.85
Tolerance	0.001	0.0005	0.001	0.0005	0.0005
Precision	0.9933	0.9947	0.9765	0.9865	0.9931
Recall	1.0	0.9995	1.0	0.9987	1.0

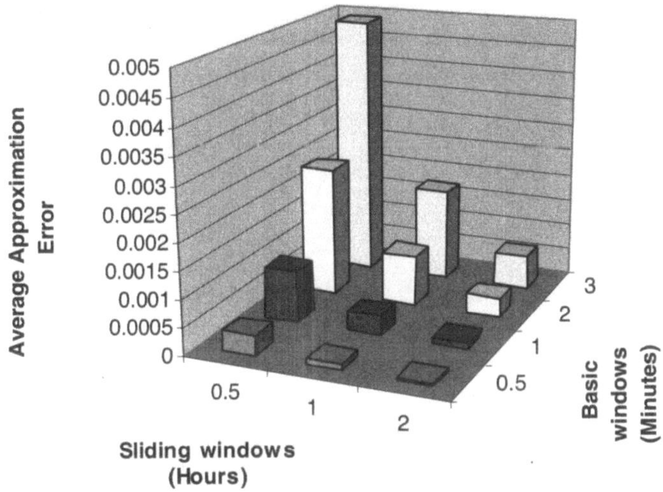

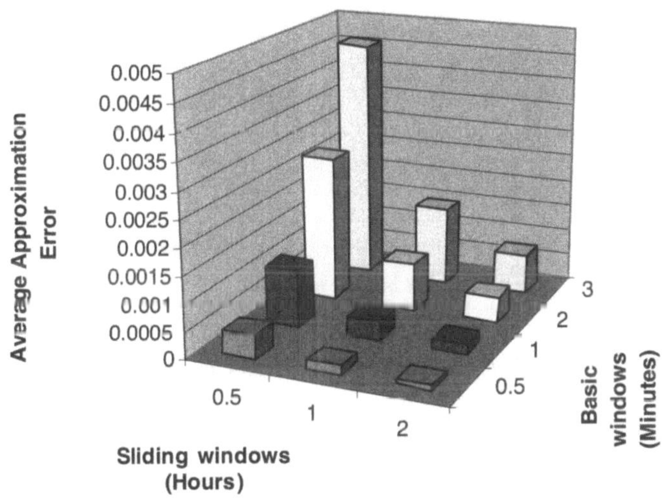

Figure 5.9. Average approximation errors for correlation coefficients with different basic / sliding window sizes for synthetic (*top*) and real (*bottom*) datasets.

5.6 Related Work

There is increasing interest in data streams. In the theoretical literature, Datar et al. [29] study the problem of maintaining data stream statistics over sliding windows. Their focus is single stream statistics. They propose an online data structure, called an exponential histogram, that can be adapted to report statistics over sliding windows at every timepoint. They achieve this with a limited memory and a tradeoff of accuracy. Our online basic window synopsis structure can report the precise single stream statistics for those timepoints that fall at basic window boundaries with a de-

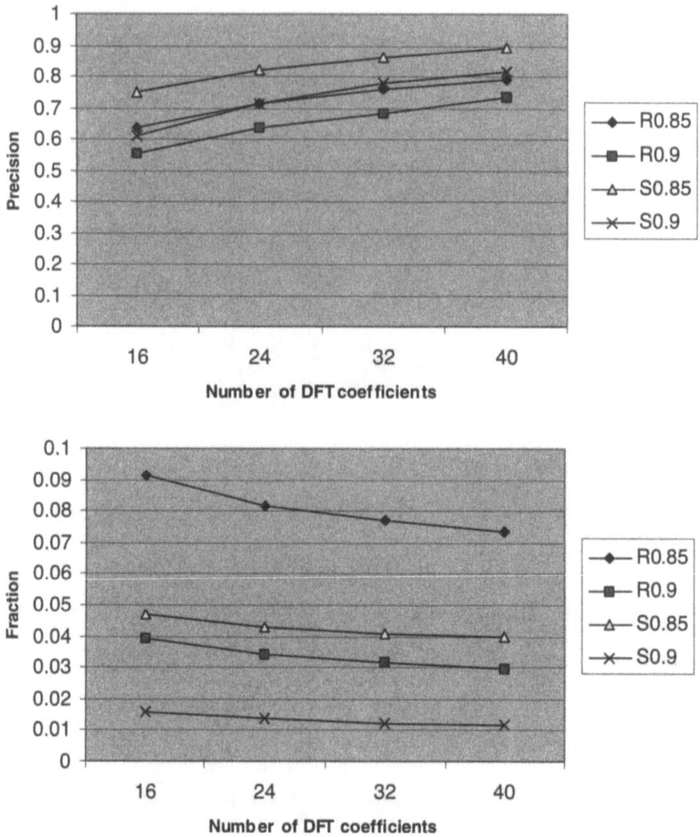

Figure 5.10. Precision and pruning power using different numbers of coefficients, thresholds and datasets.

lay of the size of a basic window, but our multi-stream statistics also trade accuracy against memory and time.

Gehrke et al. [36] also study the problem of monitoring statistics over multiple data streams. The statistics they are interested in are different from ours. They compute correlated aggregates when the number of streams to be monitored is small. A typical query in phone call record streams is the percentage of international phone calls that are longer than the average duration of a domestic phone call. They use histograms as summary data structures for the approximate computing of correlated aggregates.

Recently, the data mining community has turned its attention to data streams. A domain-specific language, Hancock [26], has been designed at AT&T to extract signatures from massive transaction streams. Algorithms for constructing decision trees [32] and clustering [42] for data streams have been proposed. Recent work of Manku et al. [68], Greenwald and Khanna [41] have focused on the problem of approximate quantile computation for individual data streams. Our work is complementary to the data mining research because our techniques for finding correlations can be used as inputs to the similarity-based clustering algorithms.

The work by Yi et al. [101] for the online data mining of co-evolving time sequences is also complementary to our work. Our approximation algorithm can spot correlations among a large number of co-evolving time sequences quickly. Their method, MUSCLES can then be applied to those highly correlated streams for linear regression-based forecasting of new values.

Time series problems in the database community have focused on discovering the similarity between an online sequence and an indexed database of previously obtained sequence information. Traditionally, the Euclidean similarity measure is used. The original work by Agrawal et al. [8] utilizes the DFT to transform data from the time domain into frequency domain and uses multidimensional index structure to index the first few DFT coefficients. In their work, the focus is on whole sequence matching. This was generalized to allow subsequence matching [34]. Rafiei and Mendelzon [81] improve this technique by allowing transformations, including shifting, scaling and moving average, on the time series before similarity queries. The distances between sequences are measured by the Euclidean distance plus the costs associated with these transformations. Our work differs from them in that (1) in [8, 34, 81, 66], the time series are all finite data sets and the focus is on similarity queries against a sequence database having a preconstructed index. Our work focuses on similarity detection in multiple online streams in real time. (2) We use the correlation coefficients as a distance measure like [66]. The correlation measure is invariant under shifting and scaling transformations. Correlation makes it possible to construct an efficient grid structure in bounded DFT feature space. This is what enables us to give real time results. [66] allows false negatives, whereas our method does not.

Other techniques such as Discrete Wavelet Transform (DWT) [21, 97, 78, 38], Singular Value Decomposition (SVD) [61] and Piecewise Constant Approximation (PCA) [99, 55] have also been proposed for similarity search as we discussed in Chapter 2. Keogh et al. [54] compares these techniques for time series similarity queries. The performance of these techniques varied depending on the characteristics of the datasets, because no single transform can be optimal on all datasets. These techniques based on curve fitting are alternative ways of computing digests and could be used in our sliding window/basic window framework.

5.7 Conclusion

Maintaining multi-stream and time-delayed statistics in a continuous online fashion is a significant challenge in data management. Statstream solves this problem in a scalable way that gives a guaranteed response time with high accuracy.

The Discrete Fourier Transform technique reduces the enormous raw data streams into a manageable synoptic data structure and gives good I/O performance. For any pair of streams, the pair-wise statistic is computed in an incremental fashion and requires constant time per update using a DFT approximation. A sliding/basic window framework is introduced to facilitate the efficient management of streaming data digests. We reduce the correlation coefficient similarity measure to a Euclidean measure and make use of a grid structure to detect correlations among thousands of high speed data streams in real time. Experiments conducted using synthetic and real data show that StatStream detects correlations efficiently and precisely.

5.8 Questions

1. The Best Fit Slope of a time series is the slope of the line that best fits the time series:

$$\mathrm{bfs}(\mathbf{s}) = \frac{\frac{1}{k}\sum_{i=1}^{k} s_i i - \mathrm{avg}(\mathbf{i})\,\mathrm{avg}(\mathbf{s})}{\frac{1}{k}\sum_{i=1}^{k} i^2 - (\mathrm{avg}(\mathbf{i}))^2}.$$

   How can you maintain the Best Fit Slope of a time series incrementally?

2. Prove Lemma 5.6.
3. To compute the first 10 DFT coefficients of a time series of size 1000, would you use the FFT routine or obtain the result directly based on the definition of DFT? Use numerical experiments to support your answer.
4. Let

$$f_m(i) = \frac{1}{\sqrt{b}} e^{j2\pi m i/b},$$

   prove that

$$\sum_{i=1}^{b} f_m(i) f_p^*(i) = \begin{cases} 0 & m \neq p \\ V(m) \neq 0 & m = p. \end{cases}$$

5. From Figure 5.6, can you estimate k_1 and k_2? Suppose we want to monitor 12,000 streams using the same computing resources, what will be the time delay?
6. Can you design a fast algorithm to compute the autocorrelation of a time series for all the possible time lags? That is, given a time series $\mathbf{x} = (x_1, x_2, \ldots, x_N)$, compute the correlation between

$$\mathbf{x}_{N-n+1} = (x_{N-n+1}, x_{N-n+2}, \ldots, x_N)$$

   and the following time series:

$$\mathbf{x}_1 = (x_1, x_2, \ldots, x_n),$$
$$\mathbf{x}_2 = (x_2, x_3, \ldots, x_{n+1}),$$
$$\ldots,$$
$$\mathbf{x}_{N-n} = (x_{N-n}, x_{N-n+1}, \ldots, x_{N-1}).$$

6

Query by Humming

Summary. The goal of a Query by Humming system is to allow a user to find a song by humming part of the tune. No musical training is needed. The problem is still unsolved. Some systems have low retrieval precision because they rely on melodic contour information from the hum tune, which in turn relies on the error-prone note segmentation process. Some systems yield better precision when matching the melody directly from audio, but they are slow because of their extensive use of Dynamic Time Warping (DTW) (see Chapter 4). HumFinder [106, 107] improves both the retrieval precision and speed compared to previous approaches. We treat music as a time series and exploit and improve well-developed techniques from time series databases to indexing the music for fast similarity queries. We improve on existing DTW indexes technique by introducing the concept of *envelope transforms*, which gives a general guideline for extending existing dimensionality reduction methods to DTW indexes. The net result is high scalability. We test our system through experiments. Please read this approach as a case study of the techniques you have seen, not as a complete solution to this hard problem.

6.1 Introduction

You have a tune lingering in your head for many days, but you don't know where you heard this tune or which song it is from. You can't search it using an online search engine such as Google because you know nothing about the metadata of the tune. Often, a music store clerk acts as a musical search engine, interpreting tunes hummed by shoppers and directing them to an album. This works even for shoppers who can't hum very well, as the authors know from personal experience. Such a resource is very useful but hard to find.

A Query by Humming system is just such a resource. The user will hum a piece of tune into a microphone connected to a computer. The computer will search a database of tunes to find a list of melodies that are most similar to the user's "query." The user will then listen to this result to see if it is actually the tune that he had in mind. Sometimes the user won't get the tune he wanted because he hummed way off tune, the database does not contain that tune, or the computer is not intelligent enough to tell whether two tunes sound similar.

Query by humming is a particular case of "Query by Content" in multimedia databases. One queries a symbolic database of melodies (such as MIDIs or digital music scores), rather than a general acoustic database (such as MP3s). Although these two formats can be linked by metadata such as artist and song names, there is no known method to extract melodies from MP3. Querying acoustic databases [98] is an interesting problem, but it is not the focus of a query by humming system. Most research into "query by humming" in the multimedia research community uses the notion of "Contour" information. Melodic contour is the sequence of relative differences in pitch between successive notes [37]. It has been shown to be a method that the listeners use to determine similarities between melodies. However, the inherent difficulty the contour method encounters is that there is no known algorithm that can reliably transcribe the user's humming to discrete notes. We discuss this point further below.

Recently, there has been work that matches a melody directly from audio [70, 50, 102]. This generally gives better query results because it is free from the error-prone note segmentation. However, because such work relies on Dynamic Time Warping (DTW), the performance is poor.

The database community has been investigating similarity query in time series databases for a long time [8, 34]. Here we will show that query by humming is a natural application for time series database research. Time series database techniques, especially dynamic time warping indexes, can be applied to build a fast and robust query by humming database system.

6.2 Related Work

Most of the research in pre-existing query by humming systems uses pitch contour to match similar melodies [37, 18, 71, 92]. The user's humming is transcribed to a sequence of discrete notes and the contour information is extracted from the notes. This contour information is represented by a few letters. For example, ("U", "D", "S") represents that a note is above, below, or the same as the previous one. Naturally, more letters can be introduced to get a finer measure of the contour. For example, "u" indicates that a note is slightly higher than the previous one while "U" indicates that it is much higher. The tunes in the databases are also represented by contour information. In this way, a piece of melody is represented by a string with a small alphabet. The edit distance can be used to measure the similarity between two melodies. Techniques for string matching such as "q-grams" can be used to speed up the similarity query. The advantage of using contours is that while most users can hum the contour correctly, they cannot hum the contour intervals correctly. But such work suffers from two serious problems.

- It has been shown that contour information alone is not enough to distinguish a large database of melodies. A typical query of six notes on a database of 2,697 tracks would result in about 330 tracks being returned [91].
- It is very hard to segment a user's humming into discrete notes. There are reliable algorithms to transcribe the user humming in each short time period to a specific

pitch [90]. But no good algorithm is known to segment such a time series of pitches into discrete notes. The precision of the query system thus rests on an imprecise preprocessing stage.

To avoid the first problem, one can use longer query lengths and finer measures of contour intervals. But that requires too much of users, especially poor singers. The second problem is more fundamental. There are two solutions proposed in the literature.

1. The user is required to clearly hum the notes of the melody using only the syllable "ta", "la" or "da" [65, 71]. But such an input method is very unnatural for users. Moreover, when there are tie notes in a tune, it is very hard for the user to articulate them correctly. Such a cumbersome job should be left to intelligent computer programs.
2. Some recent work [70, 50, 102] proposes to match the query directly from audio based on melody slope [102] or dynamic time warping [70, 50] to match the hum-query with the melodies in the music databases. The results reported using such methods are quite encouraging. Compared to note based methods, such direct methods generally have higher retrieval precision [70]. But this quality improvement comes at a price. Performance deteriorates to such an extent that [70] states: "Perhaps the biggest criticism of our work is that it is clearly a brute force approach and it is very slow." No indices are used, which makes searching in a large music database impractical.

The database community has been researching problems in similarity query for time series databases for many years. The techniques developed in the area might shed light on the query by humming problem. Agarawal et al. [8] utilized the Discrete Fourier Transform (DFT) to transform data from the time domain into the frequency domain and used a multidimensional index structure to index the first few DFT coefficients. The focus in their work was on whole sequence matching. This was generalized to allow subsequence matching [34, 74]. Rafiei and Mendelzon [81] improved this technique by allowing transformations, including shifting, scaling and moving average, on the time series before similarity queries. In addition to DFT [8, 81, 103], Discrete Wavelet Transform (DWT) [21, 97, 78], Singular Value Decomposition (SVD) [61], Piecewise Aggregate Approximation (PAA) [99, 53] and Adaptive Piecewise Constant Approximation [54] approaches have also been proposed for similarity searching.

Allowing Dynamic Time Warping (DTW) in time series similarity searching is very critical for a query by humming system. A point-by-point distance measure between time series is very likely to fail due to variations in the duration of notes. Berndt and Clifford [17] introduced the concept of DTW to the data mining community. They showed how to use Dynamic Programming to compute the DTW distance and demonstrated its application as a time series similarity measure. Yi et al. [100] were the first to investigate the DTW in very large databases. They proposed two techniques to speed up DTW in a pipeline fashion. The first technique is to use FastMap to index time series with the DTW distance measure. But this technique might result in false negatives. The second is a global lower-bounding technique for filtering out unlikely matches. In recent work, Keogh [53] proposed a technique for the exact indexing of DTW that guarantees no false negatives.

6.2.1 Insights from HumFinder

In this research, we investigate the problem of indexing large music databases, which allows efficient and effective query by humming. Our strategy and contributions are as follows.

- We treat both the melodies in the music databases and the user humming input as time series. Such an approach allows us to integrate many database indexing techniques into a query by humming system, improving the quality of such system over the traditional (contour) string databases approach.
- We design an indexing scheme that is invariant to shifting, time scaling and local time warping. This makes the system robust and allows flexible user humming input, while attaining high speeds.
- We improve on the state-of-the-art indexing technique for time series databases allowing dynamic time warping due to [53] by giving a better lower-bound distance for dynamic time warping. This yields less false negatives and shows improvements of speed by 3 to 10 times in practice.
- We formulate a general method for dimensionality reduction transforms on time series envelopes. Existing dimensionality reduction transforms, such as Discrete Fourier Transform (DFT) and Discrete Wavelet Transform (DWT), as discussed in Chapter 2, can be extended to index time series allowing dynamic time warping while avoiding false negatives. This might have applications to video processing in the spirit of [53].
- The net effect is that our approach is scalable to large music databases, as we demonstrate with our system, HumFinder.

6.3 Query by Humming System

A typical query by humming system includes three components:

- User humming: the input hum-query
- A database of music
- Indexing into the database for efficient retrieval of the hum-query

In this section, we will discuss these components in detail, with a focus on the indexing techniques, which extend those from Chapter 3.

6.3.1 User Humming: The Input Hum-query

The user hums the query melody using a PC microphone with a single channel (mono). This acoustic input is segmented into frames of 10ms and each frame is resolved into a pitch using a pitch tracking algorithm [90]. This results in a time series of the pitches. Figure 6.1 shows an example of a pitch time series. It is the tune of the first two phrases in the Beatles song "Hey Jude." The user may hum in any way he/she prefers. From the example of Figure 6.1, we can see that it is very hard for the human being to mark the borders between notes. This is also the case for the computer. As mentioned above, we are not aware of any algorithm or product that can automatically segment notes with high accuracy. That is why our system avoids note segmentation.

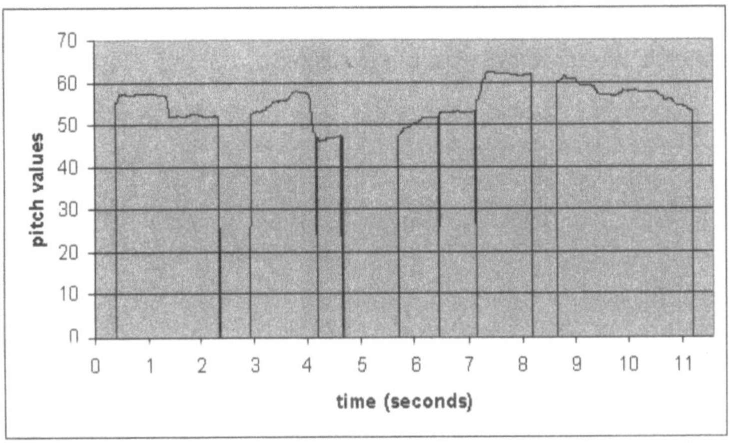

Figure 6.1. An example of a pitch time series. It is the tune of the first two phrases in the Beatles song "Hey Jude" hummed by an amateur.

6.3.2 A Database of Music

Our music database is made up of a collection of melodies. A melody is made up of a sequence of (Note, Duration) tuples. Because we use a monotone melody, there is only one note playing at each moment. A sequence of tuples (N_1, d_1), (N_2, d_2), $\ldots$, (N_k, d_k) represents a melody starting with note N_1 that lasts for d_1 time, followed by note N_2 that lasts for d_2 time $\ldots$ etc. Notice that we do not include the information of *rests* in the melody because amateur singers are notoriously bad in the timing of rests. In fact, we simply ignore the silent information in the user input humming and the candidate melodies in the database. Such a sequence of tuples can then be thought of as a time series in the following format:

$$\underbrace{N_1, N_1, \ldots, N_1}_{d_1}, \underbrace{N_2, N_2, \ldots, N_2}_{d_2}, \ldots$$

Figure 6.2 shows the melody of the beginning two phrases in the Beatles' "Hey Jude" and its time series representation.

Users are not expected to hum the whole melody. For the system to recognize sub-melodies, two methods are possible.

1. **Subsequence matching** There are many techniques for subsequence queries proposed in time series database research [34, 74], but subsequence queries are generally slower than whole sequence queries because the size of the potential candidate sequences is much larger.
2. **Whole sequence matching** We can segment each melody into several pieces based on the musical information, because most people will hum melodic sections. The query will be matched with each small piece of melody in the database.

In this research, we use whole sequence matching.

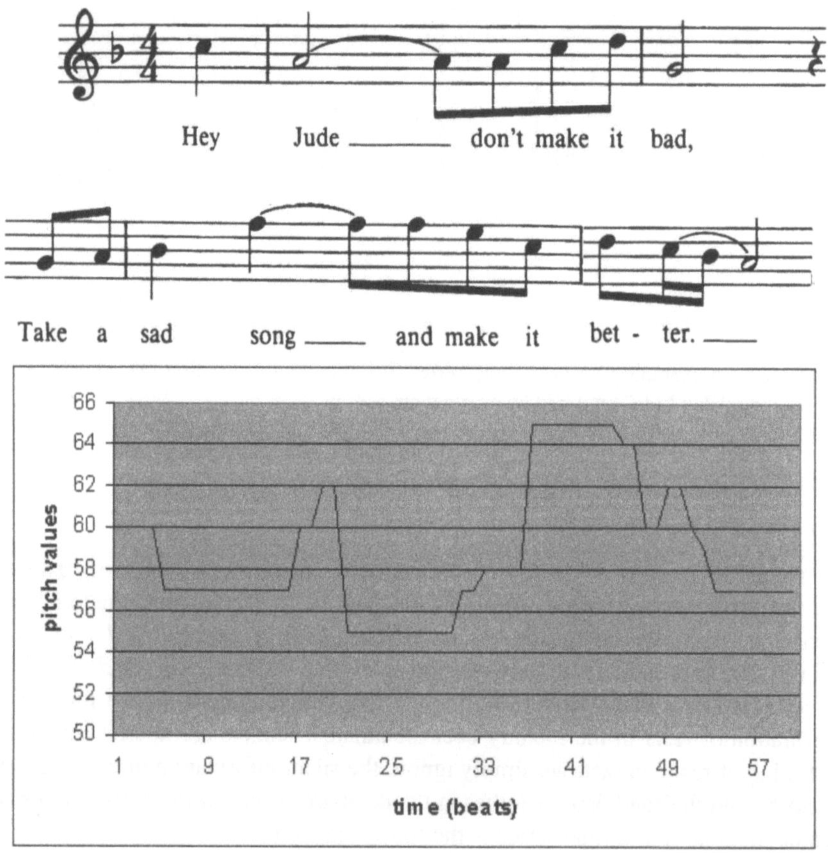

Figure 6.2. The sheet music of "Hey Jude" and its time series representation.

6.3.3 Indexing Databases for Efficient Hum Query Retrieval

If the user of the query by humming system were a good singer, we would just use the Euclidean distance between the time series to match the input pitch time series with the candidate time series in the database. The difficulty of query by humming comes from the fact that most users will not hum at the right pitch or tempo. The system should be flexible enough to allow typical inaccuracies in:

1. **Absolute pitch** Only about 1 in 10,000 people can get the absolute pitch right [80]. People of different genders, ages or even in different moods will hum the same melody with very different pitches. The humming of most people will have more accuracy in the relative pitch, or the intervals. Our system allows the user to hum at different absolute pitches. To do this, we subtract from the time series their average pitches before the matching. This is a Shift-invariant technique for time series matching.
2. **Tempo** A song can be sung at different tempos and still sound quite the same. In practice, a melody will be hummed at a tempo that ranges from half to double the original tempo. However the tempo is usually more or less consistent, that is, when the tempo changes, the duration of each note changes proportionally. We can imagine this as a uniform stretching or squeezing of the time axis. In time series database research, this is called Time Scaling, or Uniform Time Warping.
3. **Relative pitch** The problem of variation in relative pitch for the average singer is less severe than that of the absolute pitch. But it is still not rare for notes to be sung a bit high or low. Suppose that the timing of each note is perfect, the distance between the query humming time series and a candidate time series can then be measured by the sum of the differences at each sample time moment. The smaller this distance, the more similar a candidate melody is to the humming. So the problem of finding a similar melody is a Nearest Neighbor's query.
4. **Local timing variation** Unfortunately, it is not realistic to require that the timing of each humming note be perfect. Using Dynamic Time Warping, we can make the matching process allow variations in tempo for each note. The idea is to stretch and squeeze the time axis locally to minimize the point-to-point distance of two time series. An important contribution of our research is an efficient indexing scheme for local dynamic time warping.

In short, our time series approach first transforms time series to a "normal form" [39] to make the similarity measure invariant under shifting and time scaling. Figure 6.3 shows an example of the time series representations of the humming and the candidate music tune after they are transformed to their normal forms, that is, they have the same absolute pitch and tempo. The dynamic time warping distance between the normal forms of the time series will be used as the similarity measure.

To scale up to large databases, we must avoid a linear scan which examines the distance between the query time series and all the time series in the database. However, it is hard to index time series data because of their high dimension. To address the problem, the GEMINI framework [34] is used. The idea is to approximate the original time series and to reduce their dimensionality. Given a time se-

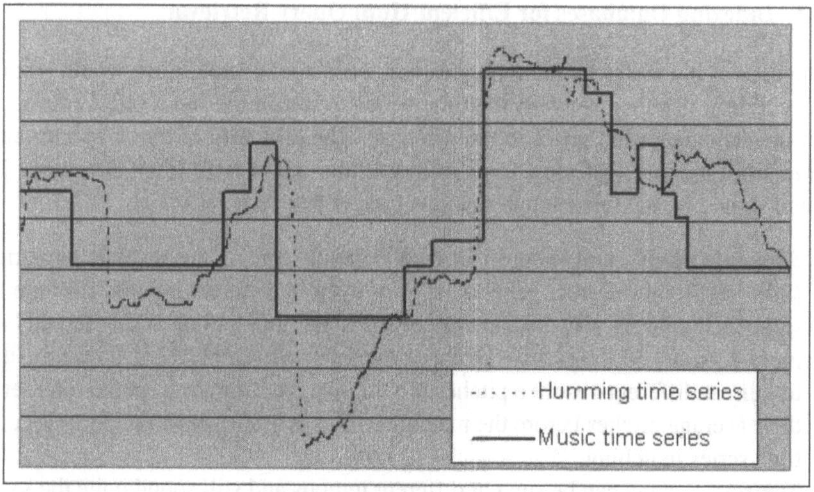

Figure 6.3. The time series representations of the hum query and the candidate music tune after they are transformed to their normal forms

ries $\mathbf{x}^n$, a dimensionality reduction transform $\mathcal{T}$ will reduce it to a lower dimension $\mathbf{X}^N = \mathcal{T}(\mathbf{x}^n), N << n$. $\mathbf{X}^N$ is also called the feature vector of $\mathbf{x}^n$. After the time series are mapped to a lower dimensionality space, they can be indexed by a multi-dimensional index structure such as an R* tree [15] or a grid file [103]. To guarantee no false negatives in similarity queries, $\mathcal{T}$ must be lower-bounding, that is, the distance between time series under dimensionality reduction should lower-bound their original distance:

$$D(\mathcal{T}(\mathbf{x}), \mathcal{T}(\mathbf{y})) \leq D(\mathbf{x}, \mathbf{y}).$$

Popular dimensionality reduction transformations include the Fourier Transform, the Wavelet Transform, SVD and Piecewise Aggregate Approximation. In the next section, we will extend the GEMINI framework to the Dynamic Time Warping distance measure.

6.4 Indexing Scheme for Dynamic Time Warping

To simplify our notation, we will first review the concept of envelope [53] for time series.

Definition 6.1 (Envelope). *The k-Envelope of a time series* $\mathbf{x} = x_i, i = 1, \ldots, n$ *is*

$$\mathrm{Env}_k(\mathbf{x}) = (\mathrm{Env}_k^L(\mathbf{x}); \mathrm{Env}_k^U(\mathbf{x})). \tag{6.1}$$

$\mathrm{Env}_k^L(\mathbf{x})$ *and* $\mathrm{Env}_k^U(\mathbf{x})$ *are the upper and lower envelope of* $\mathbf{x}$ *respectively:*

$$\mathrm{Env}_k^L(\mathbf{x}) = x_i^L, i = 1, \ldots, n; x_i^L = \min_{-k \le j \le k} (x_{i+j}) \tag{6.2}$$

$$\mathrm{Env}_k^U(\mathbf{x}) = x_i^U, i = 1, \ldots, n; x_i^U = \max_{-k \le j \le k} (x_{i+j}). \tag{6.3}$$

$\mathbf{e} = (\mathbf{e}^L; \mathbf{e}^U)$ denotes the envelope of a time series. The distance between a time series and an envelope is defined naturally as follows:

Definition 6.2. *The distance between a time series* $\mathbf{x}$ *and an envelope* $\mathbf{e}$ *is*

$$D(\mathbf{x}, \mathbf{e}) = \min_{\mathbf{z} \in \mathbf{e}} D(\mathbf{x}, \mathbf{z}). \tag{6.4}$$

We use $\mathbf{z}^n \in \mathbf{e}$ to denote that $e_i^L \le z_i \le e_i^U, i = 1, 2, \ldots, n$. So the value at each point can be any one in the range.

Keogh [53] proved that the distance between a time series and the envelope of another time series lower-bounds the true DTW distance.

Lemma 6.3. [53]

$$D(\mathbf{x}^n, \mathrm{Env}_k(\mathbf{y}^n)) \le D_{\mathrm{DTW}(k)}(\mathbf{x}^n, \mathbf{y}^n). \tag{6.5}$$

The lemma can be explained visually in Figure 6.4. The shadowed area in the figure corresponds to the distance between the time series x and the envelope of the time series y. This distance lower-bounds the true DTW distance between x and y.

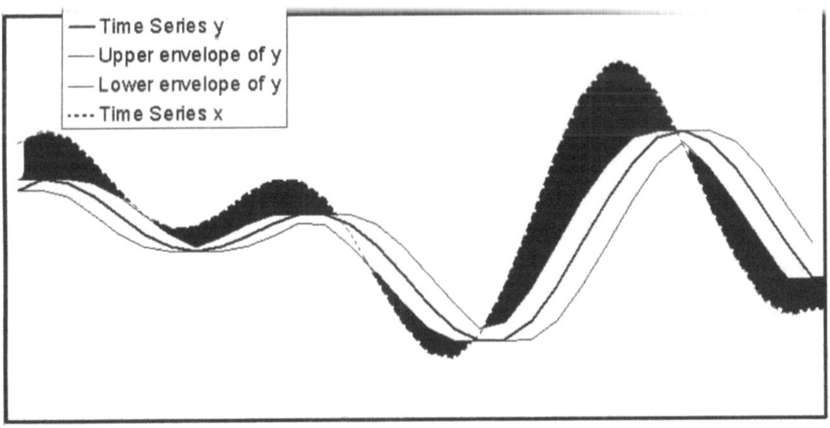

Figure 6.4. Illustration of the lower bounding Dynamic Time Warping distance.

To index the time series in the GEMINI framework, one needs to perform dimensionality reduction transform on the time series and its envelope. Piecewise Aggregate Approximation (PAA) is used in [53]. The PAA reduction of the envelopes using Keogh's method is as follows. Let $(\mathbf{L}^N; \mathbf{U}^N)$ be the PAA reduction of an envelope $(\mathbf{l}^n; \mathbf{u}^n)$,

$$L_i = \min(l_{\frac{n}{N}(i-1)+1}, \ldots, l_{\frac{n}{N}i}),$$
$$U_i = \max(u_{\frac{n}{N}(i-1)+1}, \ldots, u_{\frac{n}{N}i}), \quad i = 1, 2, \ldots, N. \tag{6.6}$$

The PAA of an envelope is just the piecewise constant function, which bounds but does not intersect the envelope.

We introduce a new PAA reduction as follows.

$$L_i = \frac{N}{n} \sum_{j=\frac{n}{N}(i-1)+1}^{\frac{n}{N}i} l_j, //U_i = \frac{N}{n} \sum_{j=\frac{n}{N}(i-1)+1}^{\frac{n}{N}i} u_j, \quad i = 1, 2, \ldots, N. \tag{6.7}$$

In our method, **U** and **L** are also piecewise constant functions, but each piece is the average of the upper or lower envelope during that time period. Figure 6.5a shows a time series, its bounding envelope and the PAA reduction of the envelope using Keogh's method. Figure 6.5b shows the same time series, its bounding envelope and the PAA reduction of the envelope using our method. We can see clearly that the bounds in Figure 6.5b are tighter than in Figure 6.5a and it is straightforward to prove that this is always the case for any time series. We will show that our bounds can still guarantee to lower-bound the real DTW distance.

Before we prove that our PAA transform can provide a lower-bound for DTW, we will first discuss general dimensionality reduction transforms on envelopes for indexing time series under the DTW distance. We define the container property of a dimensionality reduction transform for an envelope as follows.

Definition 6.4 (Container Invariant). *We say a transformation $\mathcal{T}$ for an envelope* **e** *is "container-invariant" if*

$$\forall \mathbf{x}^n \text{ if } \mathbf{x}^n \in \mathbf{e} \text{ then } \mathcal{T}(\mathbf{x}^n) \in \mathcal{T}(\mathbf{e}). \tag{6.8}$$

Just as a transform of time series that is lower-bounding can guarantee no false negatives for Euclidean distance, a transform of envelopes that is container-invariant can guarantee no false negatives for DTW distance.

Theorem 6.5. *If a transformation $\mathcal{T}$ is container-invariant and lower-bounding, then*

$$D(\mathcal{T}(\mathbf{x}), \mathcal{T}(\mathrm{Env}_k(\mathbf{y}))) \le D_{\mathrm{DTW}(k)}(\mathbf{x}, \mathbf{y}). \tag{6.9}$$

Proof. $\mathcal{T}$ is container-invariant,

$$\therefore \forall \mathbf{z}^n \text{ if } \mathbf{z}^n \in \mathbf{e} \text{ then } \mathcal{T}(\mathbf{z}^n) \in \mathcal{T}(\mathbf{e})$$
$$\therefore \{\mathbf{z}|\mathbf{z} \in \mathrm{Env}_k(\mathbf{y})\} \subseteq \{\mathbf{z}|\mathcal{T}(\mathbf{z}) \in \mathcal{T}(\mathrm{Env}_k(\mathbf{y}))\}$$
$$\therefore \min_{\{\mathbf{z}|\mathcal{T}(\mathbf{z}) \in \mathcal{T}(\mathrm{Env}_k(\mathbf{y}))\}} D(\mathbf{x}, \mathbf{z}) \le \min_{\{\mathbf{z}|\mathbf{z} \in \mathrm{Env}_k(\mathbf{y})\}} D(\mathbf{x}, \mathbf{z}).1$$

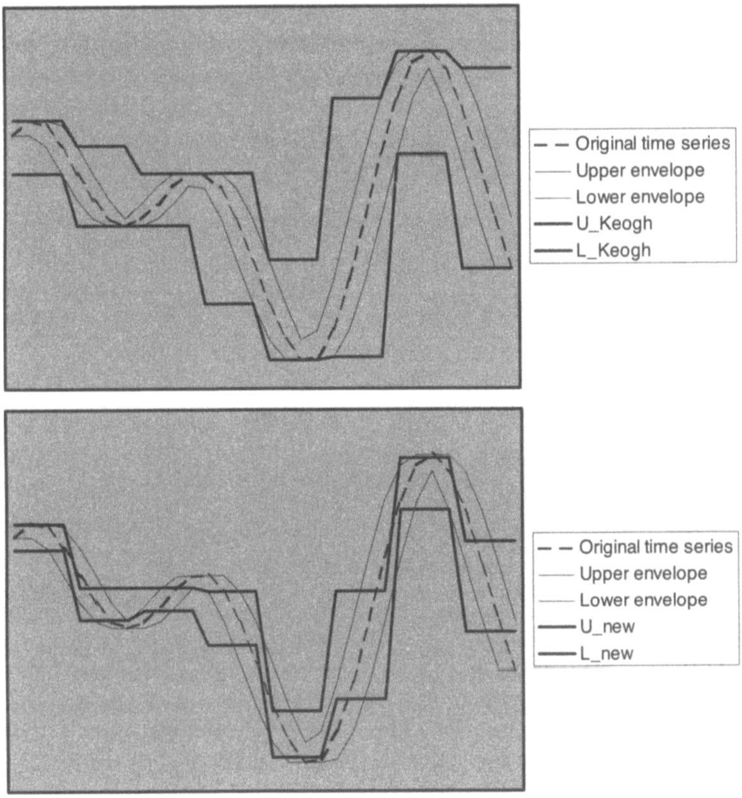

Figure 6.5. The PAA for the envelope of a time series using (a) Keogh's method (*top*) and (b) our method (*bottom*).

$\mathcal{T}$ is lower-bounding,

$$\therefore D(\mathcal{T}(\mathbf{x}), \mathcal{T}(\mathbf{z})) \leq D(\mathbf{x}, \mathbf{z})$$

$$\therefore \min_{\{\mathbf{z}|\mathcal{T}(\mathbf{z})\in\mathcal{T}(\mathrm{Env}_k(\mathbf{y}))\}} D(\mathcal{T}(\mathbf{x}), \mathcal{T}(\mathbf{z})) \leq \min_{\{\mathbf{z}|\mathcal{T}(\mathbf{z})\in\mathcal{T}(\mathrm{Env}_k(\mathbf{y}))\}} D(\mathbf{x}, \mathbf{z})$$

$$\therefore \min_{\{\mathbf{z}|\mathcal{T}(\mathbf{z})\in\mathcal{T}(\mathrm{Env}_k(\mathbf{y}))\}} D(\mathcal{T}(\mathbf{z}), \mathcal{T}(\mathbf{z})) \leq \min_{\{\mathbf{z}|\mathbf{z}\in\mathrm{Env}_k(\mathbf{y})\}} D(\mathbf{x}, \mathbf{z}).$$

By the definition of distance between time series and envelope,

$$D(\mathcal{T}(\mathbf{x}), \mathcal{T}(\mathrm{Env}_k(\mathbf{y}))) \leq D(\mathbf{x}, \mathrm{Env}_k(\mathbf{y})). \tag{6.10}$$

From Lemma 6.3,

$$D(\mathcal{T}(\mathbf{x}), \mathcal{T}(\mathrm{Env}_k(\mathbf{y}))) \leq D_{\mathrm{DTW}(k)} D(\mathbf{x}, \mathbf{y}). \qquad \square$$

Using the concept of container-invariant, we can design the transform for the envelope based on PAA, DWT, SVD and DFT. All these dimensionality reduction transforms are linear transforms[1], that is,

$$\mathbf{X}^N = \mathcal{T}(\mathbf{x}^n),$$
$$X_j = \sum_{i=1}^{n} a_{ij} x_i, \qquad j = 1, 2, \ldots, N. \tag{6.11}$$

We can extend such linear transforms for time series to transforms for time series envelopes, and at the same time guarantee that they are container-invariant.

Lemma 6.6. *Let transform $\mathcal{T}$ be a linear transform, $\mathbf{X}^N = \mathcal{T}(\mathbf{x}^n)$, $X_j = \sum_{i=1}^{n} a_{ij} x_i$, $j = 1, 2, \ldots, N$, and let the transform $\mathcal{T}$ on envelope $\mathbf{e}$ be as follows:*

$$E = (\mathbf{E}^L, \mathbf{E}^U) = \mathcal{T}(\mathbf{e}^L, \mathbf{e}^U) \tag{6.12}$$
$$E_j^U = \sum_{i=1}^{n} (a_{ij} e_i^U \tau(a_{ij}) + a_{ij} e_i^L (1 - \tau(a_{ij})))$$
$$E_j^L = \sum_{i=1}^{n} (a_{ij} e_i^L \tau(a_{ij}) + a_{ij} e_i^U (1 - \tau(a_{ij})))$$
$$j = 1, 2, \ldots, N$$

where τ is the sign function:

$$\tau(x) = \begin{cases} 1 & x \geq 0 \\ 0 & x < 0. \end{cases} \tag{6.13}$$

Then transform $\mathcal{T}$ is container-invariant.

Proof. For all $\mathbf{x}^n$, if $\mathbf{x}^n \in \mathbf{e}$ then $e_i^L \leq x_i \leq e_i^U, i = 1, 2, \ldots, n$.
Therefore $E_i^L \leq X_j \leq E_i^U, j = 1, 2, \ldots, N$,
$\therefore \ \mathcal{T}(\mathbf{x}^n) \in \mathcal{T}(\mathbf{e})$. (6.14)
□

A transform of an envelope is still an envelope: the envelope in the feature space. In the special case when the envelope equals the time series, the transform of the envelope becomes the transform of the time series. Because PAA is a linear transform, and our proposed PAA transform for envelopes is derived from the lemma above, it is container-invariant. PAA also has the nice property that all the coefficients of the linear transformation for PAA, unlike DFT and SVD, are positive. In this case, the upper envelope in feature space is just the PAA transform of the upper envelope, so is the lower envelope. For other transforms like the DFT and SVD, the upper envelope in feature space is the transform of the combination of the upper and the lower envelope. So the envelopes in the PAA feature space are tighter than those for DFT and SVD in most cases.

[1] DFT is a linear transform because the real and image part of a DFT coefficient are still a linear combination of the original time series.

Now we are ready to describe the strategy for processing hum queries with DTW. An ϵ-range similarity query in a time series database is to find all the time series whose distances with the query are less than ϵ. Here are the basic steps.

1. For each time series $\mathbf{x}^n$ in the database, compute its feature vector $\mathbf{X}^N$.
2. Build an N-dimensional index structure on $\mathbf{X}^N$.
3. For a time series query $\mathbf{q}^n$, compute its envelope $\mathbf{e}^n$ and $\mathbf{E}^N = \mathcal{T}(\mathbf{e}^n)$.
4. Make an ϵ-range query of $\mathbf{E}^N$ on the index structure, and return a set of time series S.
5. Filter out the false positives in S using their true DTW distances with $\mathbf{q}^n$. We can guarantee perfect recall because of Theorem 6.5.

Similarly, a k-nearest neighbors query can be built on top of such a range query [64, 85]. For existing time series databases indexed by DFT, DWT, PAA, SVD, etc., we can add Dynamic Time Warping support without rebuilding indices. This works because our framework can incorporate all the linear transforms and adding the DTW support requires changing only the representation of the time series query.

6.5 Software Architecture of the HumFinder System

The HumFinder system has a Graphical User Interface (GUI) to record, store and play back the user hum-query. The GUI is implemented in Java Swing and Java Sound. A snapshot of the interface is shown in Figure 6.6. The user can also browse the list of songs and listen to the music in the database through HumFinder.

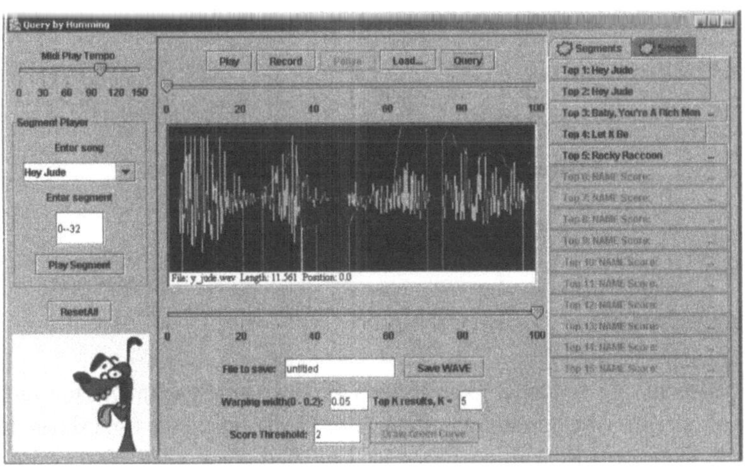

Figure 6.6. Graphical User Interface of HumFinder.

The user can hum the query melody using a PC microphone. After a hum query is recorded and transcribed to a time series of pitch values, it is sent to a server implemented in the K language. The server searches a database of melodies and returns a list of melody segments that are most similar to the hum-query. The user can listen to the playback humming when the database query is executed. The top K matches of the user hum-query will be returned. The user can listen to the results and see whether they include the target tune. The time series of the user query and the result will be displayed together. If the user doesn't get the query result he/she wants, the user can try again. As an option, the user can also change the warping width for the DTW distance and repeat the query to improve recall. Figure 6.7 shows the software architecture of the System.

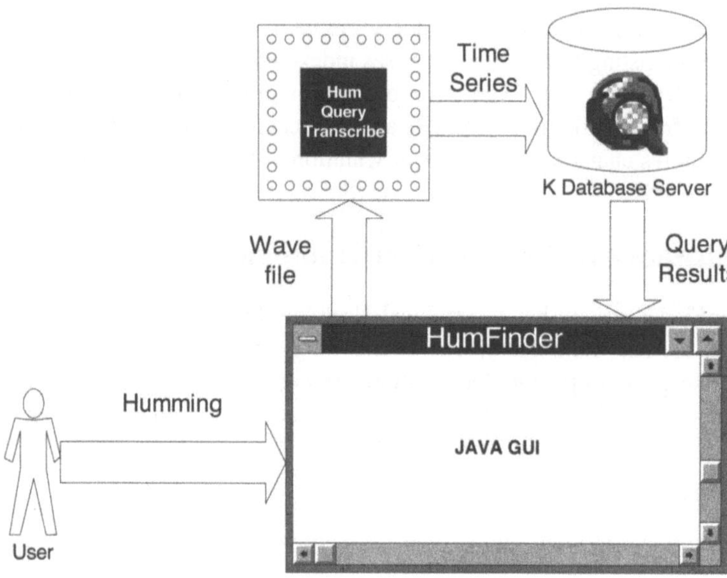

Figure 6.7. Software architecture of HumFinder.

6.6 Experiments

Our experiments are divided into three parts. First we run experiments to evaluate the quality of our query by humming system, HumFinder, using the time series database approach. Comparisons with contour-based method come next. Next, we compare the efficiency of our DTW indexing scheme to the state-of-the-art technique. Finally we will test the scalability of our system.

6.6.1 Quality of the Query by Humming System

We collected 50 of the most popular Beatles songs by manual entry. These songs are further segmented to 1000 short melodies. Each melody contains 15 to 30 notes. We asked people with different musical skills to hum for the system.

To evaluate the quality of the query by humming system, first we compared it with the note contour-based method. We will use the hum queries of better singers in this experiment, because for hum queries of poor quality it is hard for even a human being to recognize the target song. For the note contour-based method, we need to transcribe the user humming into discrete notes first. For lack of a reliable note-segmentation algorithm, we used the best commercial software we could find, *AKoff Music Composer* [9] to transcribe notes. We also applied a standard algorithm [90] to transcribe the user hum-query to a sequence of pitches, and used the silence information between pitches to segment notes. For the contour-based method, we report the better result based on these two note-segmentation processes. For the 20 pieces of hum queries by better singers, we searched the database to find their ranks using the contour-based approach and the time series approach. The result is shown in Table 6.1. We can see that our time series approach beats the contour approach clearly. We are not claiming that using contour information for music matching is bad. However until a reliable note-segmentation algorithm is developed, such an approach is based on dubious input. If for example, the query input were by piano instead of human voice so that each individual note is clearly separated, we would expect the contour-base approach to have good quality as well.

Table 6.1. The number of melodies correctly retrieved using different approaches.

Rank	Time-series Approach	Contour Approach
1	16	2
2–3	2	0
4–5	2	0
6–10	0	4
10–	0	14

We tested our system with some hum queries of poor quality, for example, by one of the authors. We define a melody to be *perfectly matched* if it is the intended target melody of the hummer and its rank is 1. The number of melodies perfectly matched is low. Still the result is quite encouraging. We noticed that the warping width can be adjusted to tune the query results. It is hard for a poor hummer to keep the right duration for each note of the melody. Allowing larger warping widths will give the hummers more flexibility in the duration of the notes. For the 20 hum queries by poor singers, we searched the database to find their ranks using DTW with different warping width. The result is reported in Table 6.2. We can see that more queries return in the top 10 matches when the warping width is increased from 0.05 to 0.1.

But this tendency disappears when the warping width is increased to 0.2, because it is unlikely for a hummer to sing way off tempo. When the warping width is too large, some melodies that are very different will have a small DTW distance too. A warping width of 1 for local DTW degenerates to global DTW. Larger warping widths also slow down the processing.

Table 6.2. The number of melodies correctly retrieved by poor singers using different warping widths.

Rank	$\delta = 0.05$	$\delta = 0.1$	$\delta = 0.2$
1	2	4	2
2–3	2	3	5
4–5	4	5	7
6–10	3	5	4
10–	9	3	2

6.6.2 Experiments on the DTW Index

Having shown that our time series approach for query by humming system compares well to the traditional contour approach, we now evaluate its efficiency. Unlike [70], the performance of the time series approach does not suffer from the extensive use of DTW. The scalability of our system comes from our proposed technique for indexing DTW. We will compare our DTW indexing technique with the best existing DTW indexing method [53] upon which we have built this work. There is an increasing awareness to use a benchmark approach in time series database experiments to guard against implementation bias and data bias. In the spirit of the work [57, 53], we took such an approach to conduct our experiments. To avoid data bias, we conducted our experiments on a wide range of time series datasets [56] that cover disciplines including finance, medicine, industry, astronomy and music. We also measured the results in an implementation free fashion to avoid bias in implementation.

We define the tightness of the lower bound for DTW distance as follows:

$$T = \frac{\text{Lower bound of DTW distance based on reduced dimension}}{\text{True DTW distance}}$$

T is in the range of [0,1]. Larger T gives a tighter bound. Note that the definition here is different from Keogh's [53]. In [53], Keogh has shown convincingly that lower-bounding using the envelope is much tighter than global bounding as reported in [100]. But such an envelope uses much more information than global lower-bounding. For a time series of size n, its envelope is represented by $2n$ values. By contrast, the global lower-bounding technique can be seen as using the minimum and maximum values of a time series as the envelope of a time series. So the global

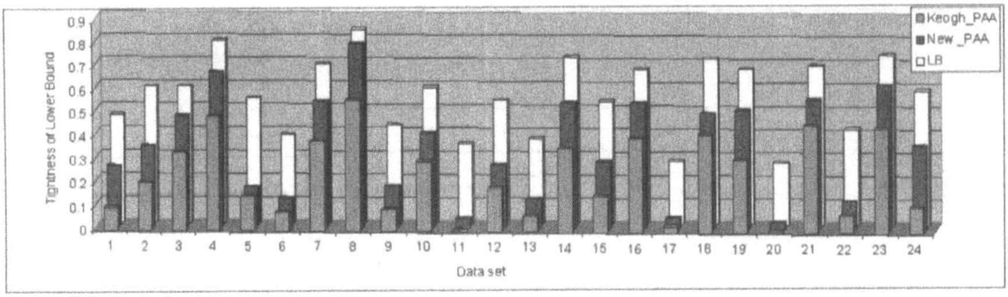

Figure 6.8. The mean value of the tightness of the lower bound, using LB, New_PAA and Keogh_PAA for different time series data sets. The data sets are 1. Sunspot; 2. Power; 3. Spot Exrates; 4. Shuttle; 5. Water; 6. Chaotic; 7. Streamgen; 8. Ocean; 9. Tide; 10. CSTR; 11. Winding; 12. Dryer2; 13. Ph Data; 14. Power Plant; 15. Balleam; 16. Standard &Poor; 17. Soil Temp; 18. Wool; 19. Infrasound; 20. EEG; 21. Koski EEG; 22. Buoy Sensor; 23. Burst; 24. Random walk

bounding requires only 2 values. To test the efficiency of dimensionality reduction under DTW, we modify the definition of T slightly, i.e., the lower-bound is based on reduced dimension. We compared three methods: LB is the lower-bound using the envelope (without dimensionality reduction and therefore without the possibility of indexing); Keogh_PAA is the lower-bound using PAA transformation proposed by Keogh [53] and New_PAA is our proposed PAA lower-bound. We chose each sequence to be of length $n = 256$ and a warping width to be 0.1. The dimension was reduced from 256 to 4 using PAA. We selected 50 time series randomly from each dataset and subtracted the mean from each time series. We computed the tightness of the lower bound of the distances between each pair of the 50 time series. The average tightnesses of the lower bound for each dataset using the three methods are reported in Figure 6.8.

From the figure, we can see that the method LB has the best T for each dataset. This is not a surprise, because the method LB uses much more information than Keogh_PAA and New_PAA. We include it here as a sanity check. LB will be used as a second filter after the indexing scheme, Keogh_PAA or New_PAA, returns a superset of answer. Using the same number of values, New_PAA is always better than Keogh_PAA. That comes from the fact that the estimations of DTW using New_PAA are always closer to the true DTW distance than the estimations using Keogh_PAA. The tightness of lower bound of New_PAA is approximately 2 times that of Keogh_PAA on average for all datasets.

This framework allows DTW indexing using different dimensionality reduction methods in addition to PAA. The performance of competing dimensionality reduction methods under the Euclidean distance measure is very data-dependent; none of them is known to be the best in all cases. Allowing different dimensionality reduction methods to be extended to the case of the DTW distance measure will provide the users with more flexibility in choosing the right method for a particular application. We tested the tightness of lower bounds for DTW indexing using dimensionality re-

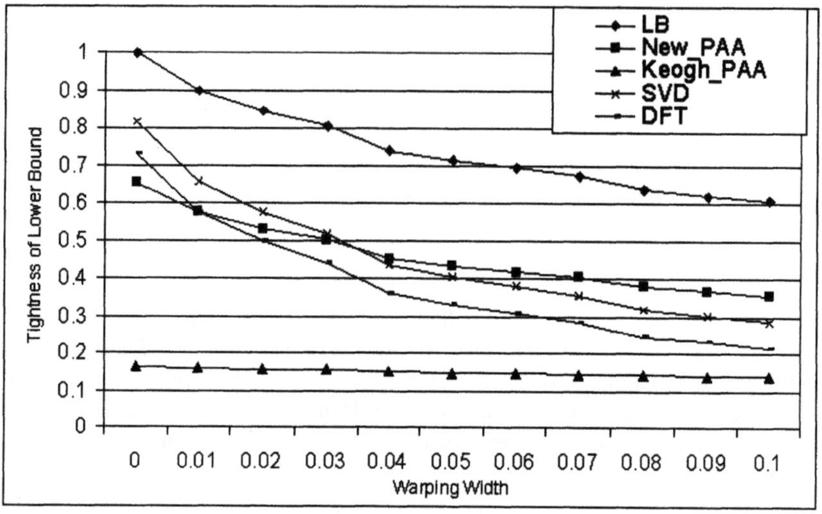

Figure 6.9. The mean value of the tightness of lower bound changes with the warping widths, using LB, New_PAA, Keogh_PAA, SVD and DFT for the random walk time series data set.

duction methods including PAA, DFT and SVD. For brevity, we will report only the results for the random walk data in Figure 6.9, because the random walk data are the most studied dataset of time series indexing and are very homogeneous. We varied the warping widths from 0 to 0.1. Each T value is the average of 500 experiments. Again LB is always the tightest lower-bound because no dimensionality reduction is performed. In the case of 0 warping width, the DTW distance is the same as the Euclidean distance. Since SVD is the optimal dimensionality reduction method for Euclidean distance, the lower-bound using SVD is tighter than any other dimensionality reduction methods. The performance of DFT and PAA for the Euclidean distance measure is similar, which confirms other research [97, 54]. For all the warping widths, New_PAA is always better that Keogh_PAA as we would expect. New_PAA also beats DFT and SVD as the warping widths increase. The reason is that all the linear transformation coefficients for PAA are positive, as we mentioned before.

6.6.3 Scalability Testing

The tightness of the lower bound is a good indicator of the performance of a DTW indexing scheme. A tighter lower-bound means that fewer candidates need to be retrieved for further examination in a particular query. That will increase the precision of retrieval at no cost to recall. Higher precision of retrieval implies lower CPU cost and IO cost at the same time, because we need to access fewer pages to retrieve candidate time series and to perform fewer exact Dynamic Time Warping computations. We will use the number of candidates retrieved and the number of page accesses as the implementation–bias-free measures for the CPU and IO cost.

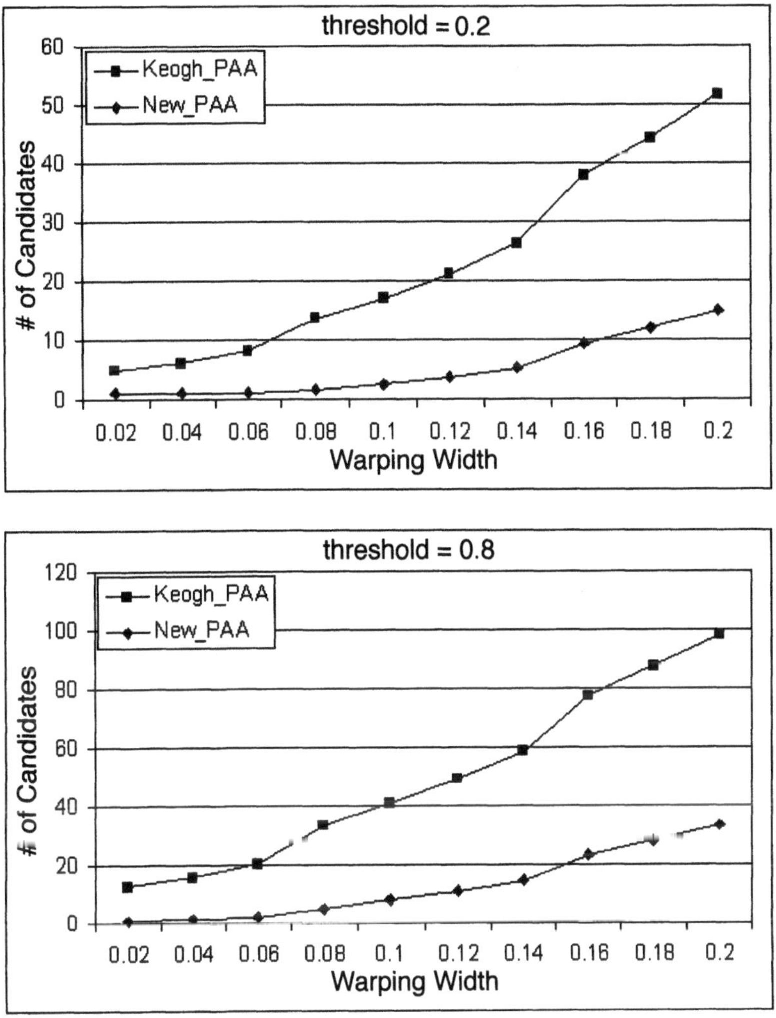

Figure 6.10. The number of candidates to be retrieved with different query thresholds for the Beatles melody database.

First we conducted experiments on our small music database of the Beatles songs. Figure 6.10 shows the average number candidates to be retrieved for queries with different selectivity. The range queries have range $n\epsilon$, n is the length of the time series and the thresholds ϵ take the values of 0.2 and 0.8. From the figure, we can see that as the warping widths get larger, the number of candidates retrieved increases, because the lower-bounds get looser for larger warping widths. Our approach (New_PAA) is up to 10 times better than Keogh_PAA.

To test the scalability of our system, we need to use larger datasets. Because of our interests, we tested a music database. We extracted notes from the melody

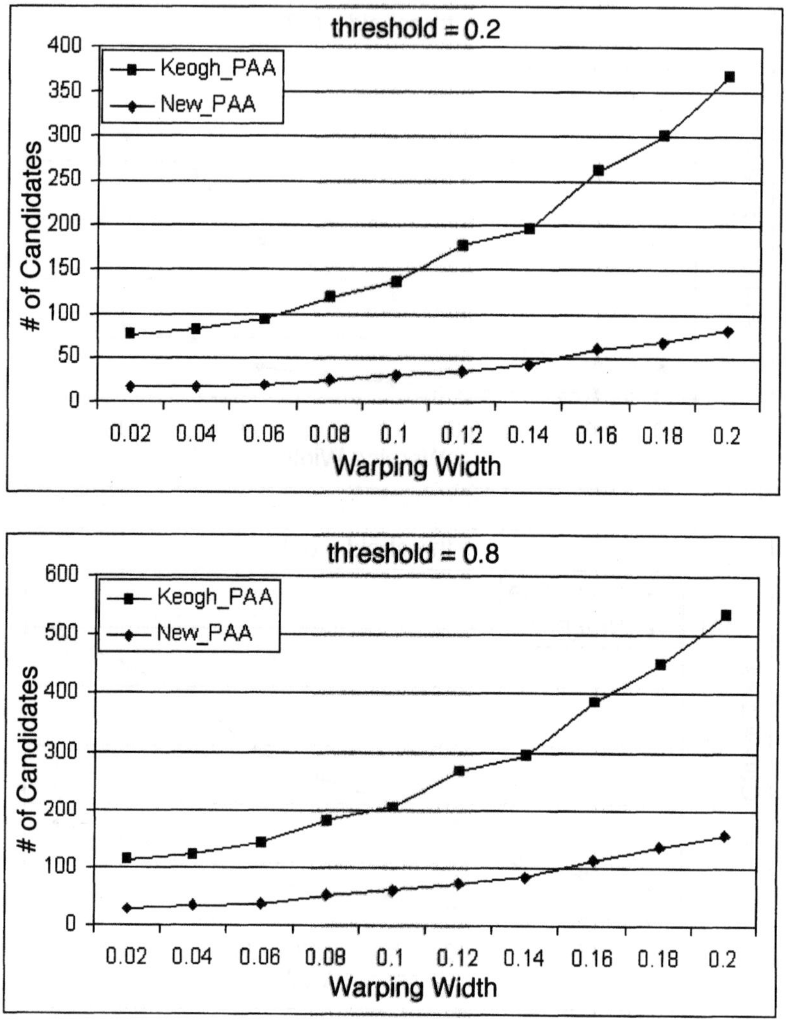

Figure 6.11. The number of candidates to be retrieved with different query thresholds for a large music database.

channel of MIDI files we collected from the Internet and transformed them to our time series representation. There are 35,000 time series in the database. The second database contains 50,000 random walk data time series. Each time series has a length of 128 and is indexed by its 8 reduced dimensions using an R* tree [15] implemented in LibGist [44]. Each result we report is averaged over 500 experiments. Figures 6.11 and 6.12 show the performance comparisons for the music database. We can see that the number of page accesses is proportional to the number of candidates retrieved for all the methods and thresholds. On a Pentium 4 PC, NEW_PAA took from 1 second for the smallest warping width to 10 seconds for the largest warping width. As the

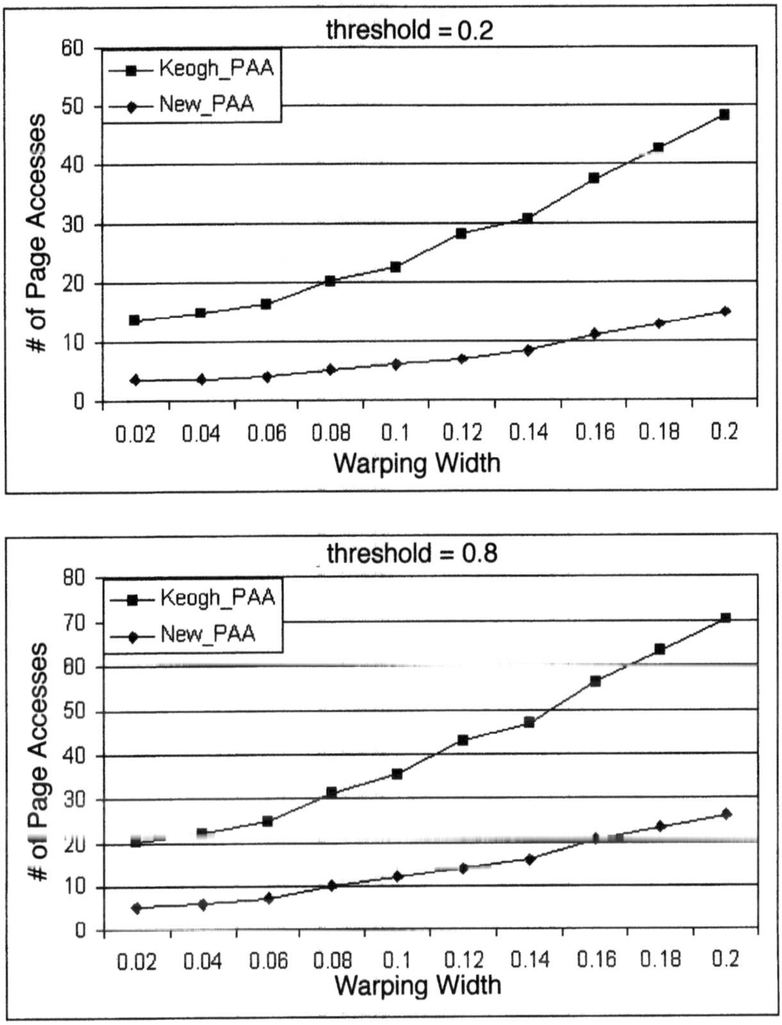

Figure 6.12. The number of page accesses with different query thresholds for a large music database.

warping width increases, the number of candidates retrieved increases significantly using the Keogh_PAA method while it increases less for New_PAA.

Figures 6.13 and 6.14 show the performance comparisons for the random walk database. Similar performance advantages of our method hold for the random walk data too.

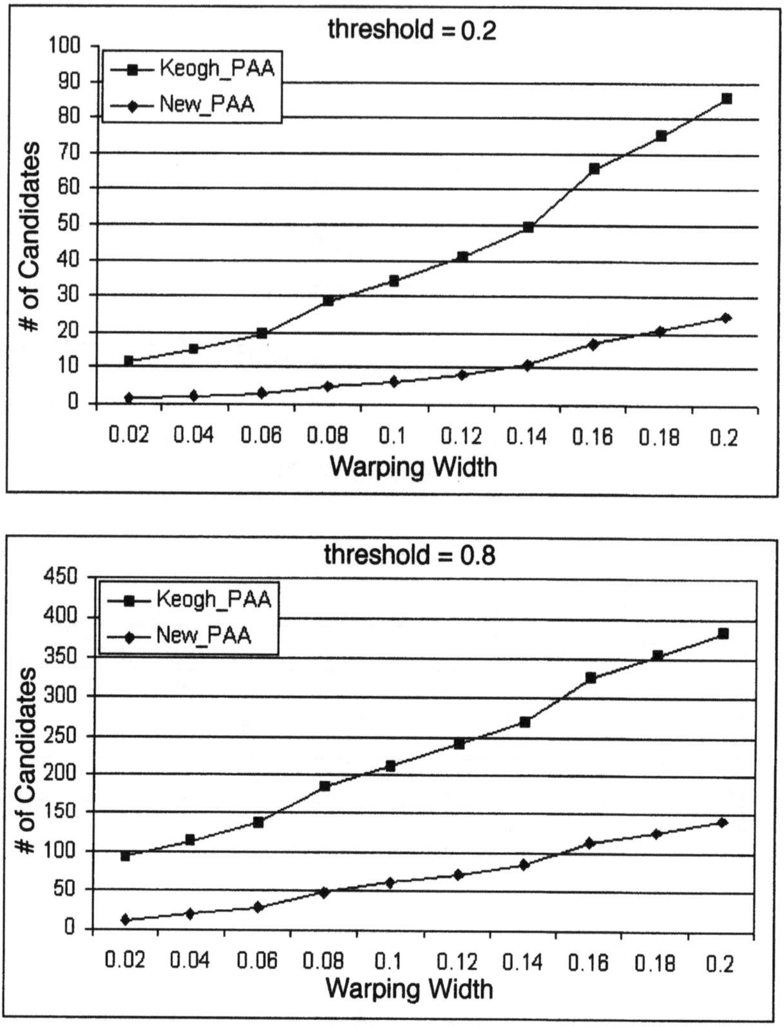

Figure 6.13. The number of candidates to be retrieved with different query thresholds for a large random walk database.

6.7 Conclusions

Besides the intrinsic interest of the application, this case study has shown that dynamic time warping lends itself well to an envelope-based index. In fact, we have shown a general approach to adapting existing time series indexing schemes for the Euclidean distance measure to the DTW distance measure. We prove that such an indexing scheme guarantees no false negatives given that the dimensionality reduction on envelope is container-invariant. Using this approach, our PAA transform for DTW is consistently better than the previous reported PAA transform. Extensive experi-

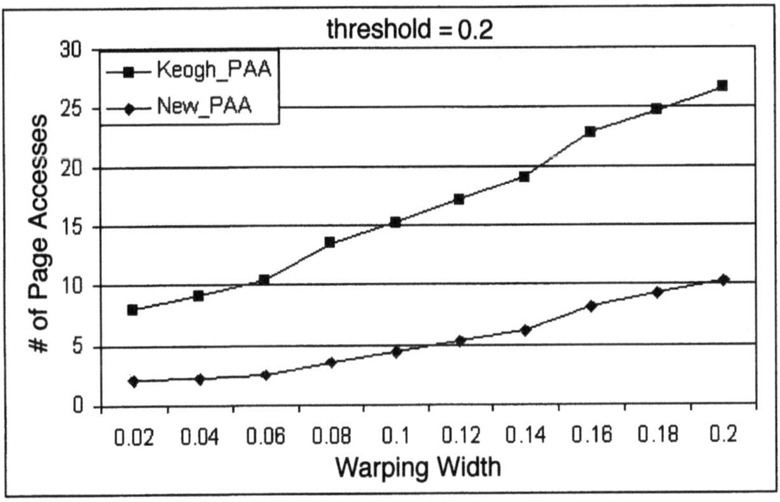

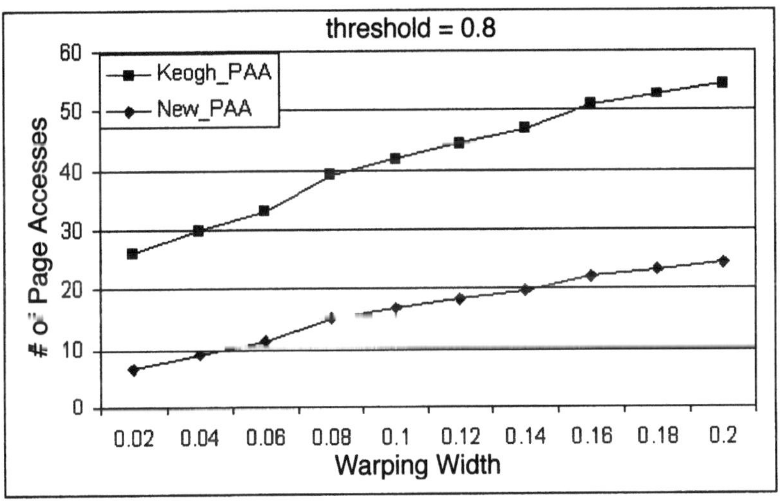

Figure 6.14. The number of page accesses with different query thresholds for a large random walk database.

ments showed that the improvement is by a factor between 3 and 10. Time warping indexes promise to give guaranteed perfect recall, high precision, and scalability.

Preliminary testing of our query by humming system HumFinder on real people (OK, our friends and a demonstration at a conference, SIGMOD 2003) gave good performance and high satisfaction. Some people even improved their singing as a result. The system is not yet mature. We are still working on expanding our melody database and adapting the system to different hummers (and even incorporating some contour information). However, the system's potential impact to entertainment and education will one day ring *fortissimo*.

6.8 Questions

1. Suppose that we live in a world in which everyone can get the tempo right. How would we be able to improve HumFinder?
2. We can use Lemma 6.6 to design a container invariant Discrete Haar Wavelet Transform for envelope. We know that the Discrete Haar Wavelet Transform and the Piecewise Aggregate Approximation (PAA) for time series are very similar. Intuitively, which one gives a tighter envelope? Try some numerical experiments to confirm or disprove that intuition.
3. Prove Lemma 6.3.
4. For Dynamic Time Warping indexing, we can either index all the time series in the database with an envelope and query them with a time series, or index the time series in the database directly and put an envelope around the query time series. We can even put an envelope around both the time series in the database and the query time series. Which approach is better? Why?
5. Given a warping width and a time series $\mathbf{x}$, we can generate all the time series that $\mathbf{x}$ can possibly "warp" to and use this set of time series to query the database using Euclidean distance. Prove that the PAA of such a set of time series is equivalent to the PAA of the envelope of $\mathbf{x}$.
6. Does the above property for the PAA hold for the DFT or the SVD as well?

7

Elastic Burst Detection

Summary. Burst detection is the activity of finding abnormal aggregates in data streams. Such aggregates are based on sliding windows over data streams. In some applications, we want to monitor many sliding window sizes simultaneously and to report those windows with aggregates significantly different from other periods. We will present a general data structure and system called OmniBurst [104] for detecting interesting aggregates over such elastic windows in near linear time. We present applications of the algorithm to detecting Gamma Ray Bursts in large-scale astrophysical data. Detection of periods with high volumes of trading activities and high stock price volatility is also demonstrated using real time Trade and Quote (TAQ) data from the New York Stock Exchange (NYSE). Our algorithm filters out periods of non-bursts in linear time, so beats the quadratic direct computation approach (of testing all window sizes individually) by several orders of magnitude.

7.1 Introduction

Consider the following application that motivates this research. An astronomical telescope, Milagro [6] was built in New Mexico by a group of prestigious astrophysicists from the Los Alamos National Laboratory and many universities. This telescope is actually an array of light-sensitive detectors covering a large pool of water about the size of a football field. It is used to continuously observe high-energy photons from the universe. When many photons are observed, the scientists assert the existence of a Gamma Ray Burst. The scientists hope to discover primordial black holes or completely new phenomena by the detection of Gamma Ray Bursts. The occurrences of Gamma Ray Bursts are highly variable, flaring on timescales of minutes to days. Once such a burst happens, it should be reported immediately. Other telescopes could then point to that portion of the sky to confirm the new astrophysical event. The data rate of the observation is extremely high. Hundreds of photons can be recorded within a second from a tiny spot in the sky [33, 48].

There are also many applications in data stream mining and monitoring when people are interested in discovering time intervals with unusually high numbers of events. For example:

- In telecommunications, a network anomaly might be indicated if the number of packages lost within a certain time period exceeds some threshold.
- In finance, stocks with unusually high trading volumes would attract the notice of the traders (or regulators) [105]. Also stocks with unusually high price fluctuations within a short time period provide more opportunity for speculation. Therefore they would be watched more closely.

Intuitively, given an aggregate function F (such as sum or count), the problem of interest is to discover subsequences s of a time series stream such that $F(s) >> F(s')$ for most subsequences s' of size $|s|$. In the case of burst detection, the aggregate is sum. If we know the duration of the time interval, we can maintain the sum over sliding windows of a known window size and sound an alarm when the moving sum is above a threshold. Unfortunately, in many cases, we cannot predict the length of the burst period. In fact, discovering that length is part of the problem to be solved. In the above example of Gamma Ray Burst detection, a burst of photons associated with a special event might last for a few milliseconds, a few hours, or even a few days. There are different thresholds associated with different durations. A burst of 10 events within 1 second could be very interesting. At the same time, a burst that lasts longer but with a lesser density of events, say 50 events within 10 seconds, could be of interest too.

Suppose that we want to detect bursts for a time series of size n and we are interested in all the n sliding window sizes. A brute-force search has to examine all the sliding window sizes and starting positions. Because there are $O(n^2)$ windows, the lower bound of the time complexity is $O(n^2)$. This is very slow for many applications. Fortunately, because we are interested in only those few windows that experience bursts, it is possible to design a nearly linear time algorithm. In this chapter we present a burst detection algorithm with time complexity approximately proportional to the size of the input plus the size of the output, i.e., the number of windows with bursts.

7.1.1 Problem Statement

There are two categories of time series data stream monitoring: *point monitoring* and *aggregate monitoring*. In point monitoring, the latest data item in the stream is of interest. When the latest item falls in some predefined domain, an alarm is sounded. For example, a stock trader who places a limited sell order on Enron informs the stock price stream monitor to raise an alarm (or automatically sell the stock) once the price of stock falls under \$10 to avoid further losses. Since only the latest data in the stream need to be considered, point monitoring can be implemented without much effort.

Aggregate monitoring is much more challenging. Aggregates of time series are computed based on certain time intervals (windows). There are three well-known window models that are the subjects of many research projects [35, 36, 38, 103, 22].

1. **Landmark windows** Aggregates are computed based on the values between a specific time point called the landmark and the present. For example, the average stock price of IBM from Jan 1, 2003 to today is based on a landmark window.
2. **Sliding windows** In a sliding window model, aggregates are computed based on the last w values in the data stream. The size of a sliding window w is predefined. For example, the running maximum stock price of IBM during the previous 5 days is based on sliding windows.
3. **Damped window** In a damped window model the weights of data decrease exponentially into the past. For example, after a new data item x is inserted, the damping moving average avg_{new} can be updated as follows:

$$\text{avg}_{\text{new}} = \text{avg}_{\text{old}} * p + x * (1 - p), 0 < p < 1.$$

The sliding window model is the most widely used in data stream monitoring. Motivated by the Gamma Ray example, we have generalized this to the *elastic window model.* In an elastic window model, the user needs to specify only the range of the sliding window sizes, the aggregate function and alarm domains, and will be notified of all those window sizes in the range with aggregates falling in the corresponding alarm domains.

Here we give the formal definition of the problem of monitoring data streams over elastic windows.

Problem 7.1. For a time series $x_1, x_2, \ldots, x_n$, given a set of window sizes $w_1, w_2, \ldots, w_m$, an aggregate function F and threshold associated with each window size, $f(w_j), j = 1, 2, \ldots, m$, monitoring elastics window aggregates of the time series is to find all the subsequences of all the window sizes such that the aggregate applied to the subsequences cross their window sizes' thresholds, i.e.,

$$\forall i \in 1..n, \forall j \in 1..m, s.t. \; F(x[i \,..\, i+w_j-1]) \geq f(w_j). \tag{7.1}$$

The threshold above can be estimated from the historical data or the model of the time series. Elastic burst detection is a special case of monitoring data streams on elastic windows. In elastic burst detection, the alarm domain is $[f(w_j), \infty)$. Note that it is also possible for the alarm domain to be $(-\infty, f(w_j)]$.

7.1.2 Insights of Omniburst

The contributions of our research are as follows:

- We introduce the concept of monitoring data streams on elastic windows and show several important applications of this model.
- We design an innovative data structure, called the Shifted Binary Tree, for efficient elastic burst monitoring. This data structure is applied to general aggregate monitoring and burst detection in higher dimensions.
- We apply our algorithm to real world data including the Milagro Gamma Ray data stream, NYSE real time tick data and text data. Our method is up to several orders of magnitudes faster than a direct computation, which means that a multi-day computation can be done in a matter of minutes.

7.2 Data Structure and Algorithm

In this section, we first give some background on the wavelet data structure, because our algorithm for burst detection is based on a similar data structure. In Section 7.2.2 we discuss the Shifted Binary Tree and the algorithm for efficient elastic burst detection in an offline setting. This is extended to a streaming algorithm in Section 7.2.3. Our algorithm will also be generalized to other problems in data stream monitoring over elastic windows in Section 7.2.4 and to higher dimensions in Section 7.2.5.

7.2.1 Wavelet Data Structure

In wavelet analysis, the wavelet coefficients of a time series are maintained in a hierarchical structure. Let us consider the simplest wavelet, the Haar wavelet. For simplicity of notation, suppose that the size of time series n is a power of two. This would not affect the generality of the results. The original time series makes up level 0 in a wavelet tree. The pair-wise (normalized) averages and differences of the adjacent data items at level 0 produce the wavelet coefficients at level 1. The process is repeated for the averages at level i to get the averages and differences at level $i+1$ until there is only one average and difference at the top level. The reader can refer to the process in computing the Haar wavelet decomposition in Table 2.2.

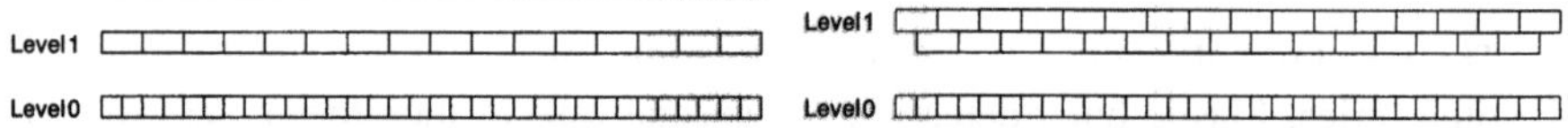

Figure 7.1. (a) Wavelet Tree (*left*) and (b) Shifted Binary Tree (*right*).

The wavelet coefficients above can also be viewed as the aggregates of the time series at different time intervals. Figure 7.1a shows the time interval hierarchy in the Haar wavelet decomposition. At level i, there are $n2^{-i}$ consecutive windows with size 2^i. All the windows at the same level are disjoint. The aggregates that the Haar wavelet maintains are the (normalized) averages and differences, while in our discussion of burst detection, the aggregate of interest is the sum. Obviously, such a wavelet tree can be constructed in $O(n)$ time.

The first few top levels of a wavelet tree yield concise multi-resolution information of the time series. This gives the wavelet lots of applications. However, for our purpose of burst detection, such a data structure has a serious problem. Because the windows at the same level are non-overlapping, a window of arbitrary start position and arbitrary size might not be included in any window in the wavelet tree, except the window at the highest level that includes everything. For example, the window consisting of three time points in the middle, $(n/2 - 1, n/2, n/2 + 1)$, is not contained in any window in the wavelet tree except the largest one. This makes wavelets inconvenient for the discovery of properties of arbitrary subsequences.

7.2.2 Shifted Binary Tree

In a *shifted binary tree* (SBT) (Figure 7.1b), the adjacent windows of the same level are half overlapping. In Figure 7.1, we can see that the size of a SBT is approximately double that of a wavelet tree, because at each level, there is an additional "line" of windows. These additional windows provide valuable overlapping information for the time series. They can be maintained either explicitly or implicitly. If we keep only the aggregates for a traditional wavelet data structure, the aggregates of the overlapping windows at level i can be computed from the aggregates at level $i - 1$ of the wavelet data structure.

To build an SBT, we start from the original time series and compute the pair wise aggregate (sum) for each two consecutive data items. This will produce the aggregates at level 1. A downsampling on this level will produce the input for the higher level in the SBT. Downsampling is simply sampling every second item in the series of aggregates. In Figure 7.1b, downsampling will retain the upper consecutive non-overlapping windows at each level. This process is repeated until we reach the level where a single window includes every data point. Figure 7.2 gives a pseudocode to build a SBT. Like regular wavelet trees, the SBT can also be constructed in $O(n)$ time.

```
Given: x[1..n],n=2^a
Return: shifted binary tree SBT[1..a][1..]

b=x;
FOR i = 1 TO a //remember a = log2 n
  //merge consecutive windows and form
  //level i of the shifted binary tree
  FOR j = 1 TO size(b)-1 STEP 2
    SBT[i][j]=b[j]+b[j+1];
  ENDFOR
  //downsample, retain a non-overlapping cover
  FOR j = 1 TO size(SBT[i])/2
    b[j]=SBT[i][2*j-1];
  ENDFOR
ENDFOR
```

Figure 7.2. Algorithm to construct a shifted binary tree.

For a subsequence starting and ending at arbitrary positions, there is always a window in the SBT that tightly includes the subsequence as Figure 7.3 shows and the following lemma proves.

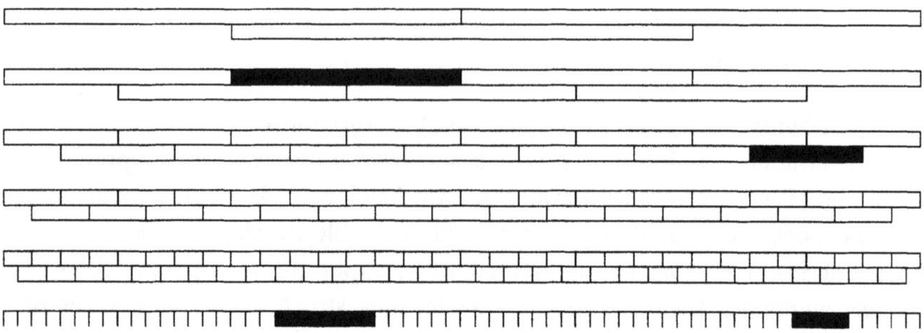

Figure 7.3. Examples of the windows that include subsequences in the shifted binary tree.

Lemma 7.2. *Given a time series of length n and its shifted binary tree, any subsequence of length* w, $w \leq 1 + 2^i$ *is included in one of the windows at level* $i + 1$ *of the shifted binary tree.*

Proof. The windows at level $i + 1$ of the shifted binary tree are

$$[(j-1)2^i+1 \,..\, (j+1)2^i],\ j = 1, 2, \ldots, \frac{n}{2^i} - 1. \tag{7.2}$$

A subsequence with length $1+2^i$ starting from an arbitrary position c will be included in at least one of the above windows, because

$$[c \,..\, c+2^i] \subseteq [(j-1)2^i+1 \,..\, (j+1)2^i],\ j = \lfloor \frac{c-1}{2^i} \rfloor + 1. \tag{7.3}$$

Any subsequence with length w, $w \leq 1 + 2^i$ is included in some subsequence(s) with length $1 + 2^i$, and therefore is included in one of the windows at level $i + 1$. We say that windows with size w, $1 + 2^{i-1} < w \leq 1 + 2^i$, are monitored by level $i + 1$ of the SBT. □

Because, for time series of non-negative numbers, the aggregate sum is monotonically increasing, the sum of the time series within a sliding window of any size is bounded by the sum of its including window in the shifted binary tree. This fact can be used as a filter to eliminate those subsequences whose sums are far below their thresholds.

Figure 7.4 gives the pseudo-code for spotting potential subsequences of size w with sums above its threshold $f(w)$, where $1 + 2^{i-1} < w \leq 1 + 2^i$. We know that the bursts for sliding window sizes in $(1 + 2^{i-1}, 1 + 2^i)$ are monitored by the level $i + 1$ in the SBT. The minimum of the thresholds of these window sizes is

```
Given:
  time series x[1..n],
  shifted binary tree at level i+1, SBT[i+1][1..],
  a set of window sizes 1+2^(i-1) < w_1 < w_2 < ... < w_m <= 1+2^i,
  thresholds f(w_k), k = 1,2,...,m
Return:
  Subsequences of size w_k, k = 1,2,...,m with burst

FOR j = 1 TO size(SBT[i+1])
  IF (SBT[i+1][j] >= f[w_1] )
  //possible burst in subsequence x[(j-1)2^i +1 .. (j+1)2^i],
    Let w_t be the window size such that
      f(w_t) <= SBT[i+1][j] < f(w_(t+1))
    //w_t is the largest window size with possible burst.
    detailed search with window size w_k, k = 1,2,...,t in
    subsequence x[(j-1)2^i +1 .. (j+1)2^i]
  ENDIF
ENDFOR
```

Figure 7.4. Algorithm to search for bursts.

$f(w_1)$ because the thresholds increase with the window sizes. If $SBT[i+1][j]$ is less than $f(w_1)$, we know that there won't be any bursts for windows monitored by level $i+1$ at position j. That is, no burst exists in the subsequence $x[(j-1)2^i + 1 \,..\, (j+1)2^i]$. If $SBT[i+1][j]$ is larger than $f(w_1)$, then there might be bursts in the subsequence. Suppose that $f(w_t)$ is the largest window size whose threshold is exceeded by $SBT[i+1][j]$, that is, $f(w_t) \leq SBT[i+1][j] < f(w_{t+1})$. We know that in subsequence $x[(j-1)2^i + 1 \,..\, (j+1)2^i]$, there might be bursts of window sizes $w_1, w_2, \ldots, w_t$. The *detailed search* for bursts of window sizes $w_1, w_2, \ldots, w_t$ on the subsequences is then performed. A detailed search of window size w on a subsequence is to compute the moving sums with window size w in the subsequence directly and to test whether these moving sums cross the burst threshold.

In the spirit of the original work of [8] that uses lower bound techniques for fast time series similarity search, we have the following lemma that guarantees the correctness of our algorithm.

Lemma 7.3. *The above algorithm can guarantee no false negatives in elastic burst detection from a time series of non-negative numbers.*

Proof. From Lemma 7.2, any subsequence of length w, $w \leq 1 + 2^i$ is contained within a window in the SBT:

$$[c \,..\, c+w-1] \subseteq [c \,..\, c+2^i] \subseteq [(j-1)2^i+1 \,..\, (j+1)2^i]. \tag{7.4}$$

Because the sum of the time series of non-negative numbers is monotonic increasing, we have

$$\sum(x[c..c+w-1]) \leq \sum(x[c..c+2^i-1]) \leq \sum(x[(j-1)2^i+1..(j+1)2^i]). \quad (7.5)$$

By eliminating sequences with lengths larger than w but with sums less than $f(w)$, we do not introduce false negatives because

$$\sum(x[(j-1)2^i+1 \,..\, (j+1)2^i]) < f(w) \Rightarrow \sum(x[c \,..\, c+w-1]) < f(w). \quad (7.6)$$

□

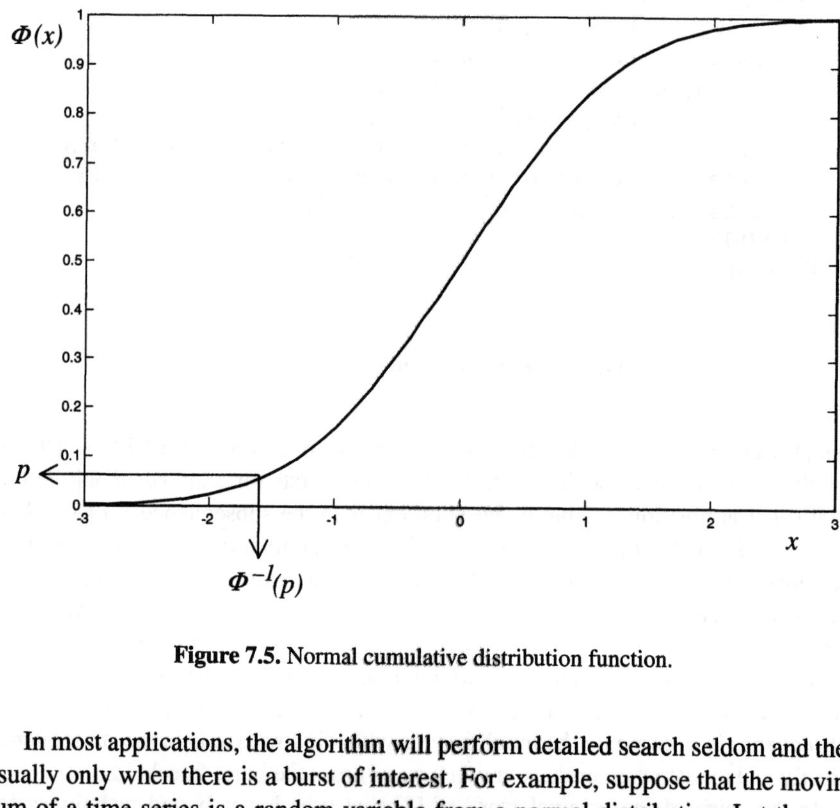

Figure 7.5. Normal cumulative distribution function.

In most applications, the algorithm will perform detailed search seldom and then usually only when there is a burst of interest. For example, suppose that the moving sum of a time series is a random variable from a normal distribution. Let the sum within sliding window w be $S_o(w)$ and its expectation be $S_e(w)$. We assume that

$$\frac{S_o(w) - S_e(w)}{S_e(w)} \sim \text{Norm}(0, 1). \quad (7.7)$$

We set the threshold of burst $f(w)$ for window size w such that the probability that the observed sums exceed the threshold is less than p, i.e., $Pr[S_o(w) \geq f(w)] \leq p$. Let $\Phi(x)$ be the normal cumulative distribution function; remember that for a normal random variable X,

$$\begin{aligned} Pr[X \leq \Phi^{-1}(p)] &\leq p \\ Pr[X \geq -\Phi^{-1}(p)] &\leq p. \end{aligned} \quad (7.8)$$

This is illustrated in Figure 7.5. We have

$$Pr\left[\frac{S_o(w) - S_e(w)}{S_e(w)} \geq -\Phi^{-1}(p)\right] \leq p. \tag{7.9}$$

Therefore,

$$f(w) = S_e(w)(1 - \Phi^{-1}(p)). \tag{7.10}$$

Because our algorithm monitors the burst based on windows with size $W = Tw$, $1 \leq T < 2$, the detailed search will report all real bursts. Actually our algorithm performs a detailed search whenever there are more than $f(w)$ events in a window of W. Therefore the rate of detailed search p_f is higher than the rate of true alarms. Suppose that $S_e(W) = TS_e(w)$; we have

$$\frac{S_o(W) - S_e(W)}{S_e(W)} \sim \text{Norm}(0, 1). \tag{7.11}$$

$$\begin{aligned} p_f &= Pr[S_o(W) \geq f(w)] \\ &= Pr\left[\frac{S_o(W) - S_e(W)}{S_e(W)} \geq \frac{f(w) - S_e(W)}{S_e(W)}\right] \\ &= \Phi\left(-\frac{f(w) - S_e(W)}{S_e(W)}\right) \\ &= \Phi\left(1 - \frac{f(w)}{TS_e(w)}\right) \\ &= \Phi\left(1 - \frac{1 - \Phi^{-1}(p)}{T}\right). \end{aligned} \tag{7.12}$$

The rate of detailed search is very small for small p, the primary case of interest. For example, let $p = 10^{-6}$, $T = 1.5$, p_f is 0.006. In this model, the upper bound of the false alarm rate is guaranteed to be small.

The time for a detailed search in a subsequence of size W is $O(W)$. The total time for all detailed searches is linear in the number of false alarms and true alarms (the output size k). The number of false alarms depends on the data and the setting of thresholds, and it is approximately proportional to the output size k. So the total time for detailed searches is bounded by $O(k)$. To build the SBT takes time $O(n)$, the total time complexity of our algorithm is approximately $O(n+k)$, which is linear in the total of the input and output sizes.

7.2.3 Streaming Algorithm

The SBT data structure in the previous section can also be used to support a streaming algorithm for elastic burst detection. Suppose that the set of window sizes in the elastic window model are $2^L < w_1 < w_2 < \cdots < w_m \leq 2^U$. For simplicity of explanation, assume that new data becomes available at every time unit.

Without the use of SBT, a naive implementation of elastic burst detection has to maintain the m sums over the sliding windows. When a new data item becomes available, for each sliding window, the new data item is added to the sum and the corresponding expiring data item of the sliding window is subtracted from the sum. The running sums are then checked against the monitoring thresholds. This takes time $O(m)$ for each insertion of new data. The response time is one time unit if enough computing resources are available.

By comparison, the streaming algorithms based on the SBT data structure will be much more efficient. For the set of window sizes $2^L < w_1 < w_2 < \cdots < w_m \leq 2^U$, we need to maintain the levels from $L+2$ to $U+1$ of the SBT that monitor those windows. There are two methods that provide tradeoffs between throughput and response time.

- **Online Algorithm** The online algorithm will have a response time of one time unit. In the SBT data structure, each data item is covered by two windows in each level. Whenever a new data item becomes available, we will update those $2(U-L)$ aggregates of the windows in the SBT immediately. Associated with each level, there is a minimum threshold. For level i, the minimum threshold δ_i is the min of the thresholds of all the windows monitored by level i, that is, $\delta_i = \min f(w_j), 2^{i-2} < w_j \leq 2^{i-1}$. If the sum in the most recently completed window at level i exceeds δ_i, it is possible one of the windows monitored by level i exceeds its threshold. We will perform a detailed search on those time intervals. Otherwise, the data stream monitor awaits insertions into the data stream. This online algorithm provides a response time of one time unit, and each insertion of the data stream requires $2(U-L)$ updates plus possible detailed searching.
- **Batch Algorithm** The batch algorithm will be lazy in updating the SBT. Remember that the aggregates at level i can be computed from the aggregates at level $i-1$. If we maintain aggregates at an extra level of consecutive windows with size 2^{L+1}, the aggregates at levels from $L+2$ to $U+1$ can be computed in batch. The aggregate in the most recently completed window of the extra level is updated every time unit. An aggregate of a window at the upper levels in the SBT will not be computed until all the data in that window are available. Once an aggregate at a certain upper level is updated, we also check alarms for time intervals monitored by that level. A batch algorithm gives higher throughput although with longer response time (with guaranteed bound close to the window size whose threshold was exceeded) than an online algorithm as the following lemmas state.

Lemma 7.4. *The amortized processing time per insertion into the data stream for a batch algorithm is* 2.

Proof. At level i, $L+2 \leq i \leq U+1$, of the SBT, every 2^{i-1} time units there is a window in which all the data are available. The aggregates at that window can be computed in time $O(1)$ based on the aggregates at level $i-1$. Therefore the amortized update time for level i is $\frac{1}{2^{i-1}}$. The total amortized update time for all levels (including the extra level) is

$$1 + \sum_{i=L+2}^{U+1} \frac{1}{2^{i-1}} \leq 2. \tag{7.13}$$

□

Lemma 7.5. *The burst activity of a window with size w will be reported with a delay less than $2^{\lceil(\log_2 w)-1\rceil}$.*

Proof. A window with size w, $2^{i-1} < w \leq 2^i$, is monitored by level $i+1$ of the SBT. The aggregates of windows at level $i+1$ are updated every 2^i time units. When the aggregates of windows at level $i+1$ are updated, the burst activity of window with size w can be checked. So the response time is less than $2^i = 2^{\lceil \log_2 w \rceil}$. □

7.2.4 Other Aggregates

It should be clear that in addition to sum, the monitoring of many other aggregates based on elastic windows could benefit from our data structure, as long as the following conditions holds.

1. The aggregate F is monotonically increasing or decreasing with respect to the window, i.e., if window $[a_1..b_1]$ is contained in window $[a_2..b_2]$, then $F(x[a_1..b_1]) \leq F(x[a_2..b_2])$ (or $F(x[a_1..b_1]) \geq F(x[a_2..b_2])$) always holds.
2. The alarm domain is one sided, that is, [threshold, ∞) for monotonically increasing aggregates and ($-\infty$, threshold] for monotonically decreasing aggregates.

The most important and widely used aggregates are all monotonic: *Max* and *Count* are monotonically increasing and *Min* is monotonically decreasing. Another monotonic aggregate is *Spread*. Spread measures the volatility or movement level of time series. Spread of a time series $\mathbf{x}$ is

$$\text{Spread}(\mathbf{x}) = \text{Max}(\mathbf{x}) - \text{Min}(\mathbf{x}). \tag{7.14}$$

Spread is monotonically increasing. The spread within a small time interval is less than or equal to that within a larger time interval. A large spread within a small time interval is of interest in many applications in data stream because it indicates that the time series has experienced large movement.

7.2.5 Extension to Two Dimensions

The one-dimensional shifted binary tree for time series can naturally be extended to higher dimensions, such as spatial dimensions. In this section we consider the problem of discovering elastic spatial bursts using a two-dimensional shifted binary tree. Given a fixed image of scattering dots, we want to find the regions of the image with an unexpectedly high density of dots. In an image of the sky with many dots representing the stars, such regions might indicate galaxies or supernovas. The problem is to report the positions of spatial sliding windows (rectangle regions) having different sizes, within which the density exceeds some predefined threshold.

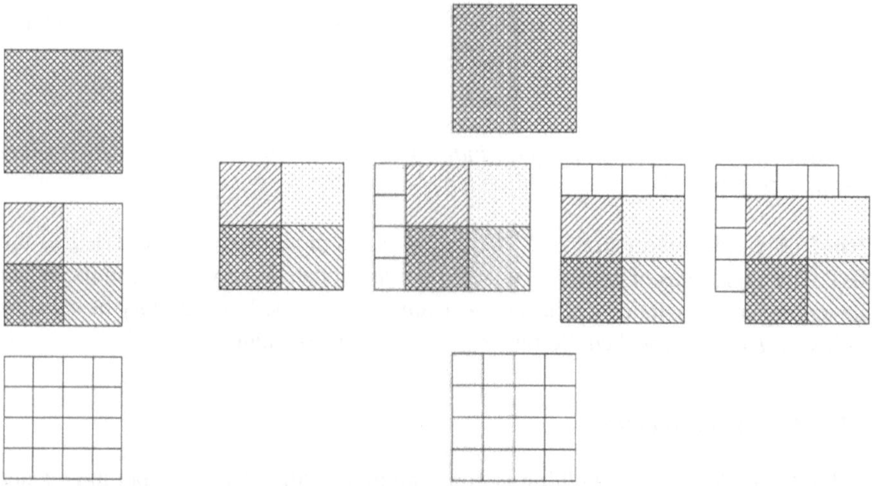

Figure 7.6. (a) Wavelet Tree (*left*) and (b) Shifted Binary Tree (*right*).

The two-dimensional shifted binary tree is based on the two-dimensional wavelet structure. The basic wavelet structure separates a two-dimensional space into a hierarchy of windows as shown in Figure 7.6a (similar to quadtree [83]). Aggregate information will be computed recursively based on those windows to get a compact representation of the data. Our two-dimensional shifted binary tree will extend the wavelet tree in a similar fashion as in the one-dimensional case. This is demonstrated in Figure 7.6b. At the same level of the wavelet tree, in addition to the group of disjoint windows that are the same as in the wavelet tree, there are another three groups of disjoint windows. One group of windows offsets the original group in the horizontal direction, one in the vertical direction and the third one in both directions.

Any square spatial sliding window with size $w \times w$ is included in one window of the two-dimensional SBT. The size of such a window is at most $2w \times 2w$. Using the techniques of Section 7.2.2, burst detection based on the SBT-2D can report all the high density regions efficiently.

7.3 Software Architecture of the OmniBurst System

OmniBurst also runs in the interpreted environment K [1]. Figure 7.7 shows the software architecture of the System. The OmniBurst system reads the data stream and updates the Shifted Binary Tree. The system also reads the configuration file that specifies the thresholds of burst for different window sizes. Based on the thresholds and the data digests in the Shifted Binary Tree, OmniBurst will report the Elastic Bursts. The thresholds are computed based on some statistical models or the historical data. These thresholds can be updated as more time series data are observed.

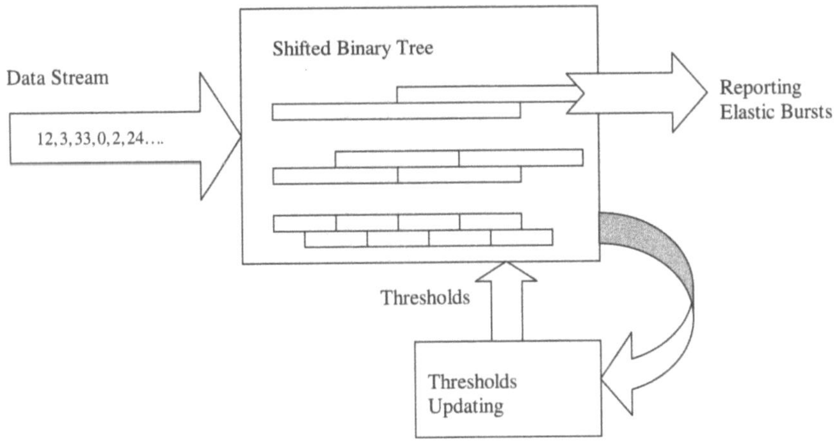

Figure 7.7. Software architecture of OmniBurst.

The readers can find the download and running instructions for OmniBurst on the website: *http://www.cs.nyu.edu/cs/faculty/shasha/papers/burst.d/burst.html*

7.4 Empirical Results of the OmniBurst System

Our empirical study will first demonstrate the desirability of elastic burst detection for some applications. We also study the performance of our algorithm by comparing our algorithm with the brute force searching algorithm in Section 7.4.2.

7.4.1 Effectiveness Study

As an emotive example, we monitor bursts of interest in countries from the presidential State of the Union addresses from 1913 to 2003. The same example was used by Kleinberg [60] to show the bursty structure of text streams. In Figure 7.8 we show the number of times that some countries were mentioned in the speeches. There are clearly bursts of interest in certain countries. An interesting observation here is that these bursts have different durations, varying from years to decades.

The rationale behind elastic burst detection is that a predefined sliding window size for data stream aggregate monitoring is insufficient in many applications. The same data aggregated in different time scales will give very different pictures as we can see in Figure 7.9. In it we show the moving sums of the number of events for about an hour's worth of Gamma Ray data. The sizes of sliding windows are 0.1, 1 and 10 seconds respectively. For better visualization, we show only those positions with bursts. Naturally, bursts at small time scales that are extremely high will produce bursts at larger time scales too. More interestingly, bursts at large time scales are not necessarily reflected at smaller time scales, because those bursts at large time scales

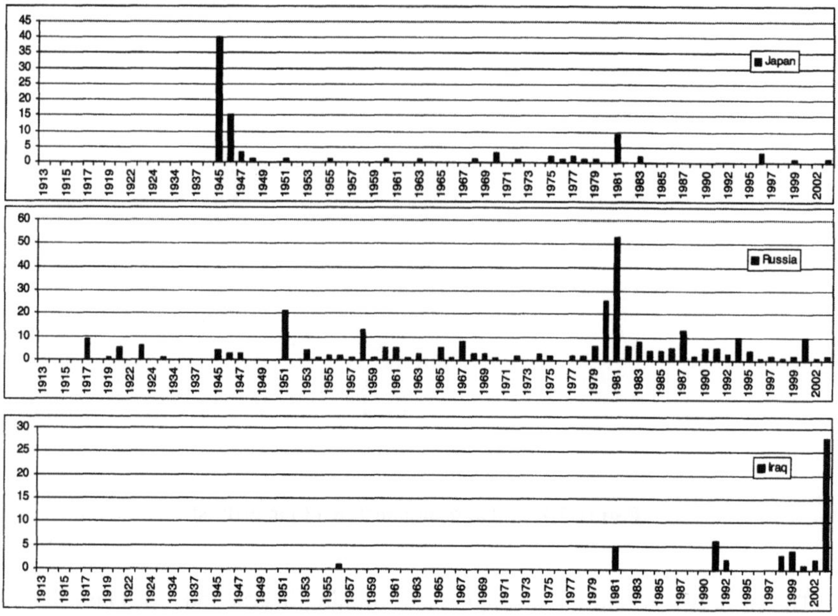

Figure 7.8. Bursts of the number of times that countries were mentioned in the President's State of the Union address.

might be composed of many consecutive "bumps." Bumps are those positions where the numbers of events are high but not high enough to be "bursts". Therefore, by looking at different time scales at the same time, elastic burst detection will give more insight into the data stream.

We also show in Figure 7.10 an example of spatial elastic bursts. We use the 1990 census data of the population in the continental United States. The populations in the map are aggregated in a grid of $0.2° \times 0.2°$ in Latitude/Longitude. We compute the total population within sliding spatial windows with sizes $1° \times 1°$, $2° \times 2°$ and $5° \times 5°$. Those regions with population above the 98 percentile in different scales are highlighted. We can see that the different sizes of sliding windows give the distribution of high population regions at different scales.

7.4.2 Performance Study

Our experiments on OmniBurst were performed on a 1.5 GHz Pentium 4 PC with 512 MB of main memory running Windows 2000. We tested the algorithm with two different types of data sets:

- The Gamma Ray data set: This data set includes 12 hours of data from a small region of the sky, where Gamma Ray bursts were actually reported during that time. The data are time series of the number of photons observed (events) every 0.1 second. There are totally 19,015 events in this time series of length 432,000.

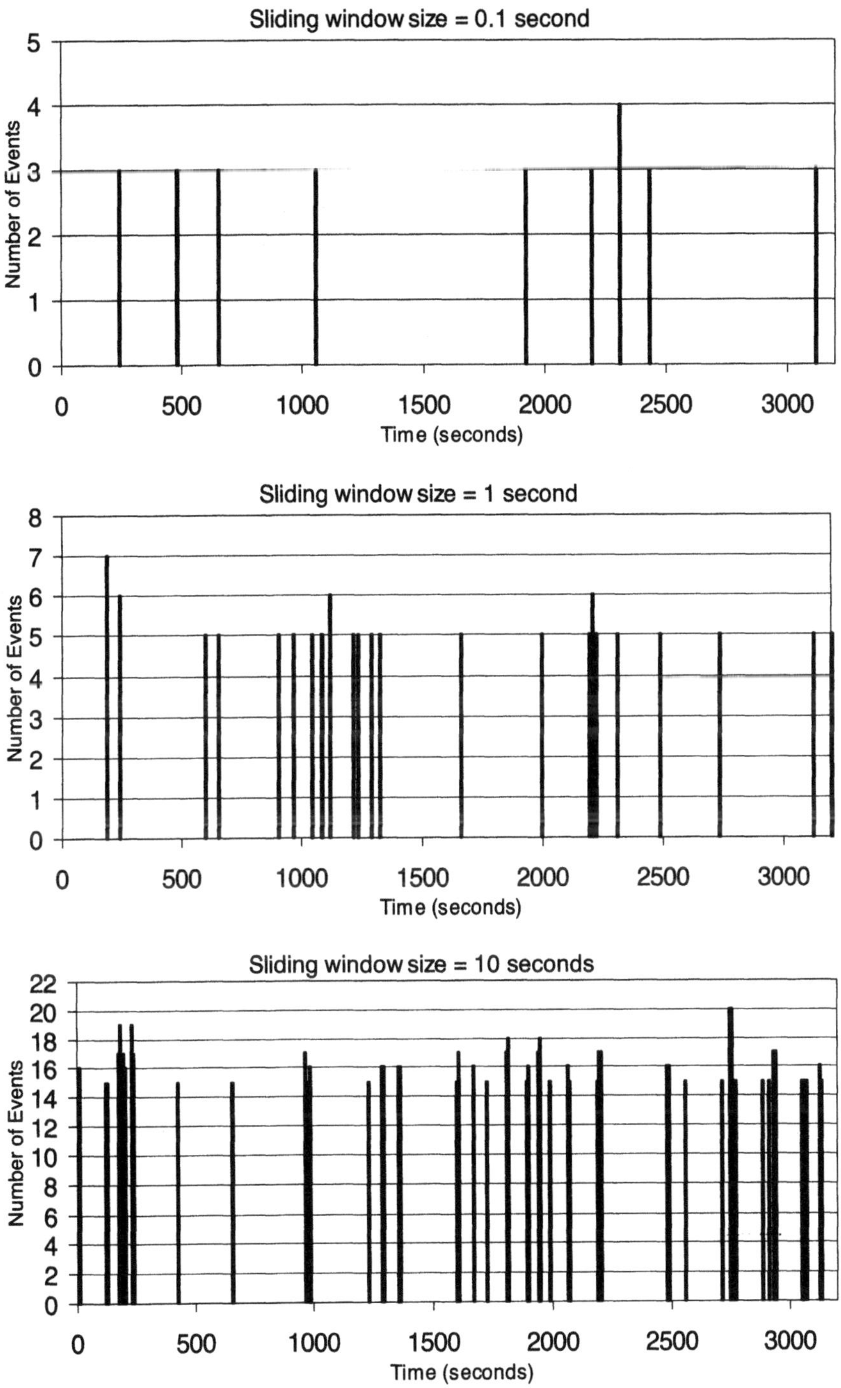

Figure 7.9. Bursts in Gamma Ray data for different sliding window sizes.

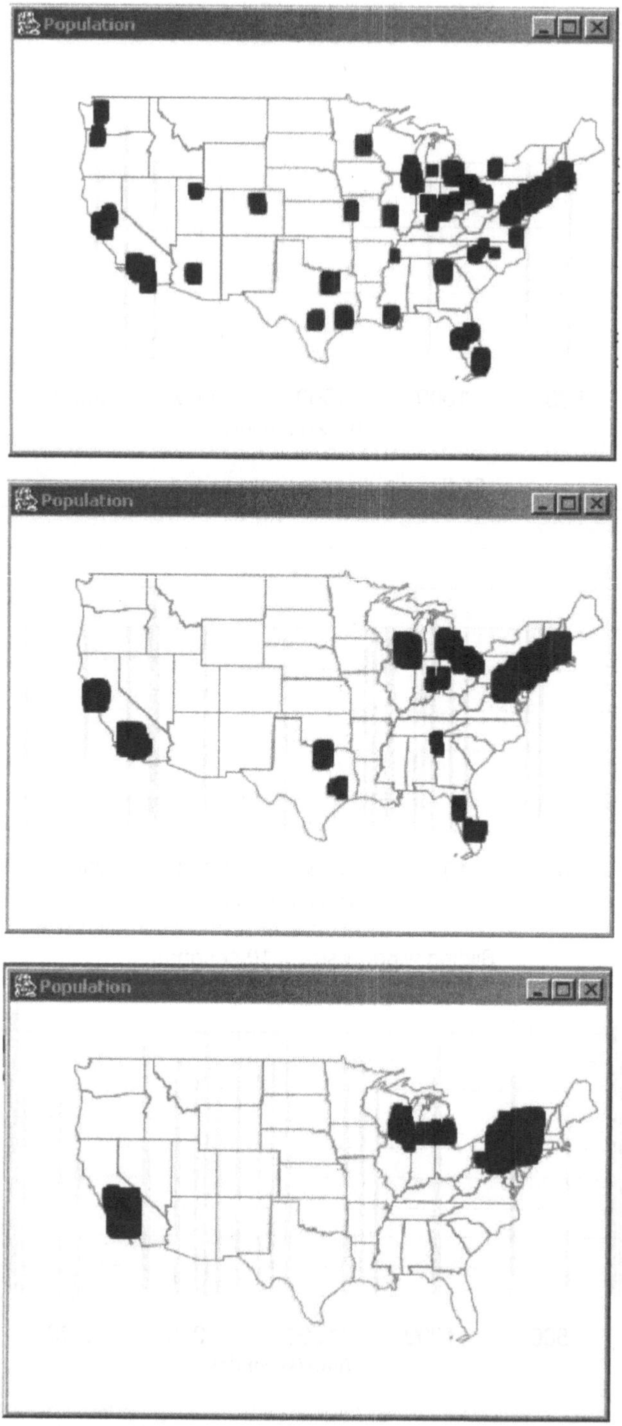

Figure 7.10. Bursts in population distribution data for different spatial sliding window sizes.

- The NYSE TAQ Stock data set: This data set includes four years of tick-by-tick trading activities of the IBM stock between July 1st, 1998 and July 1st, 2002. There are 5,331,145 trading records (ticks) in total. Each record contains trading time (precise to the second), trading price and trading volume.

In thc following experiments, we set the thresholds of different window sizes as follows. We use the first few hours of Gamma Ray data and the first year of Stock data as training data respectively. For a window of size w, we compute the aggregates on the training data with a sliding window of size w. This gives another time series $\mathbf{y}$. The thresholds are set to be $f(w) = \text{avg}(\mathbf{y}) + \xi std(\mathbf{y})$, where $\text{avg}(\mathbf{y})$ and $std(\mathbf{y})$ are the average and standard deviation respectively. The factor of threshold ξ is set to 8. The list of window sizes is $5, 10, \ldots, 5 * N_w$ time units, where N_w is the number of windows. N_w varies from 5 to 50. The time unit is 0.1 seconds for the Gamma Ray data and 1 minute for the stock data.

First we compare the wall clock processing time of elastic burst detection from the Gamma Ray data in Figure 7.11. Our algorithm based on the SBT data structure is more than ten times faster than the direct algorithm. The advantage of using our data structure becomes more obvious as we examine more window sizes. The processing time of our algorithm is output-dependent. This is confirmed in Figure 7.12, where we examine the relationship between the processing time using our algorithm and the number of alarms. Naturally the number of alarms increases as we examine more window sizes. We also observed that the processing time follows the number of alarms well. Recall that the processing time of the SBT algorithm includes two parts: building the SBT and the detailed searching of those potential portions of burst. Building the SBT takes only 200 milliseconds for the data set, which is negligible when compared to the time to do the detailed search. Also for demonstration purposes, we intentionally, to our disadvantage, set the thresholds lower and therefore got many more alarms than what physicists are interested in. If the alarms are scarce, as is the case of Gamma Ray burst detection, our algorithm will be even faster. In Figure 7.13 we fix the number of windows to be 25 and change the factor of threshold ξ. The larger ξ is, the higher the thresholds are, and therefore the fewer alarms will be sounded. Because our algorithm is dependent on the output sizes, the higher the thresholds are, the faster the algorithm runs. In contrast, the processing time of the direct algorithm does not decrease with higher thresholds.

For the next experiments, we test the elastic burst detection algorithm on the IBM Stock trading volume data. Figure 7.14 shows that our algorithm is up to 100 times faster than a brute force method. We also zoom in to show the processing time for different output sizes in Figure 7.15.

In addition to elastic burst detection, our SBT data structure works for other elastic aggregate monitoring too. In the following experiments, we search for big spreads on the IBM Stock data. Figures 7.16 and 7.17 confirm the performance advantages of our algorithm. Note that for the aggregates of Min and Max, and thus Spread, there is no known deterministic algorithm to update the aggregates over sliding windows incrementally in constant time. The filtering property of SBT data structure gains more by avoiding unnecessary detailed searching. So in this case our algorithm is up

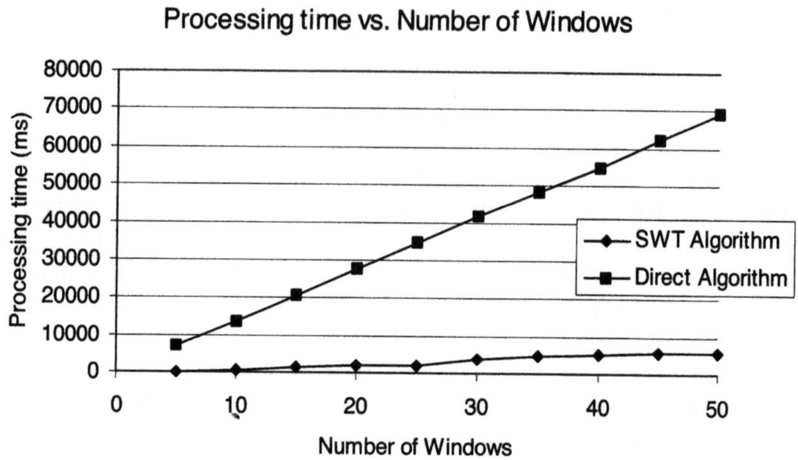

Figure 7.11. Processing time of elastic burst detection on Gamma Ray data for different numbers of windows.

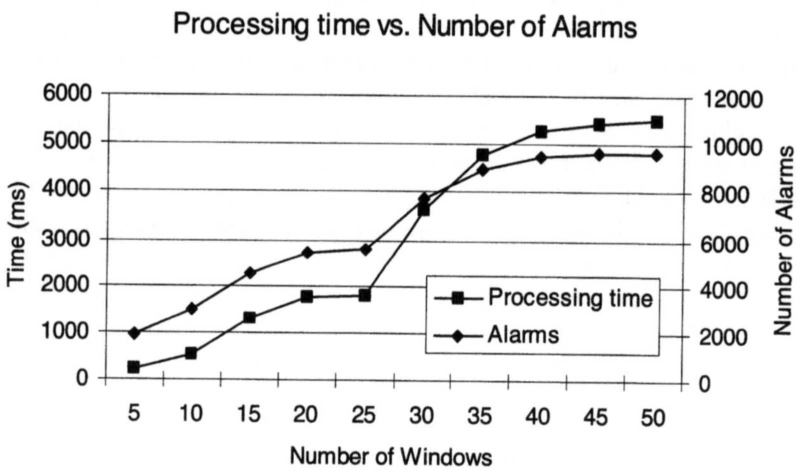

Figure 7.12. Processing time of elastic burst detection on Gamma Ray data for different output sizes.

to 1,000 times faster than the direct method, reflecting the advantage of a near linear algorithm as compared with a quadratic one.

7.5 Related work

There is much recent interest in data stream mining and monitoring. Excellent surveys of models and algorithms in data stream can be found in [11, 75]. The sliding window is recognized as an important model for data streams. Based on the sliding

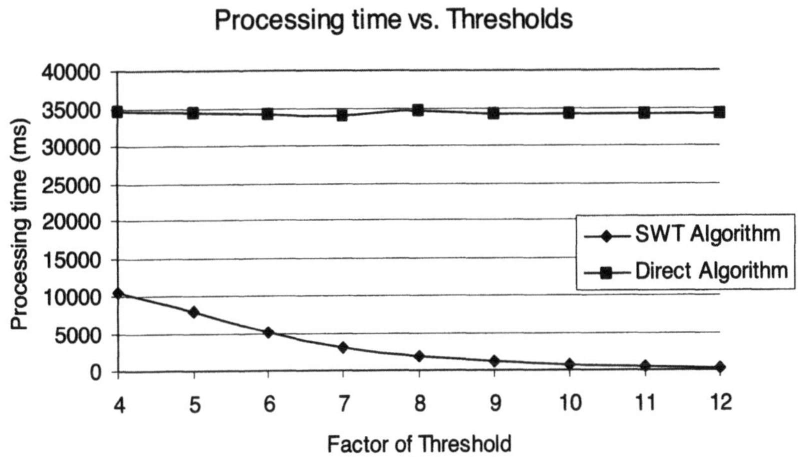

Figure 7.13. Processing time of elastic burst detection on Gamma Ray data for different thresholds.

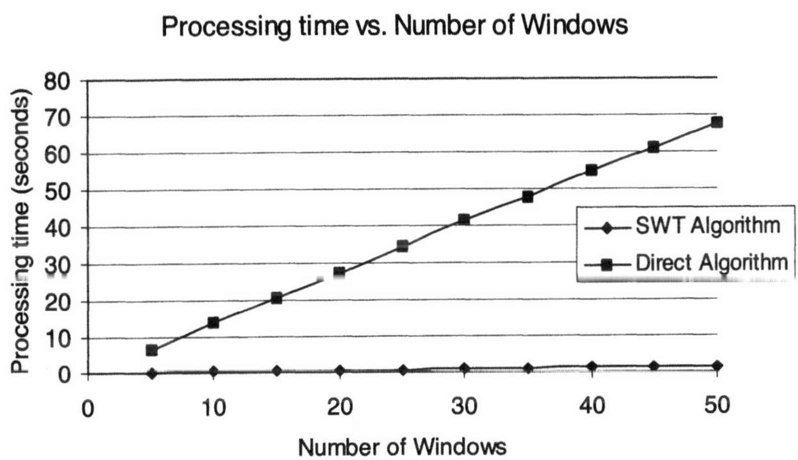

Figure 7.14. Processing time of elastic burst detection on Stock data for different numbers of windows.

window model, previous research studies the computation of different aggregates on data streams, for example, correlated aggregates [36], count and other aggregates [29], frequent itemsets and clusters [35], variance and k-medians [12], and correlation [103]. The work [45] studies the problem of learning models from time-changing streams without explicitly applying the sliding window model. The Aurora project [19] and STREAM project [10] consider the systems aspect of monitoring data streams. Also the algorithm issues in time series stream statistics monitoring are addressed in StatStream [103]. In this research we extend the sliding window

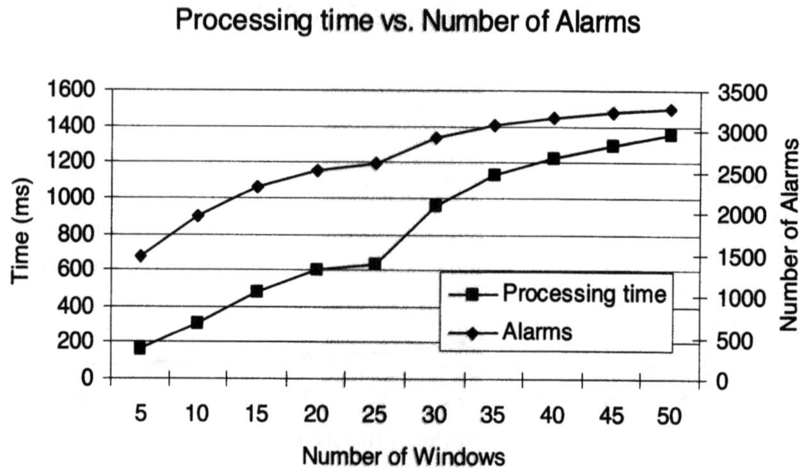

Figure 7.15. Processing time of elastic burst detection on Stock data for different output sizes.

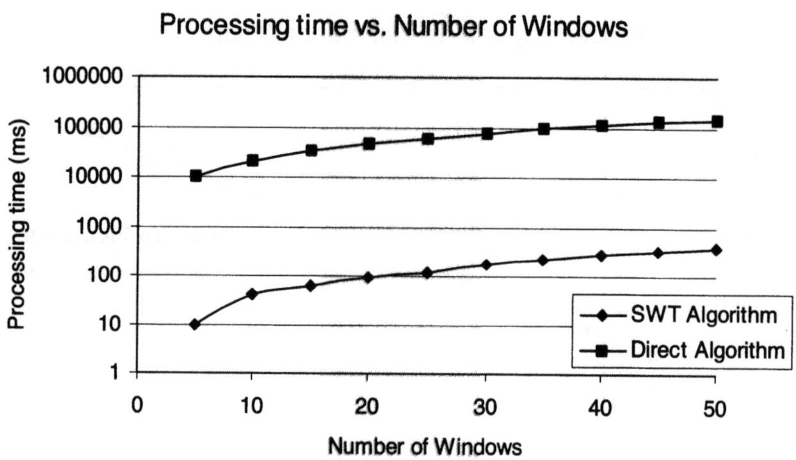

Figure 7.16. Processing time of elastic spread detection on Stock data for different numbers of windows.

model to the elastic sliding window model, making the choice of sliding window size more automatically.

Wavelets are heavily used in the context of data management and data mining, including selectivity estimation [69], approximate query processing [93, 20], dimensionality reduction [21] and streaming data analysis [38]. However, its use in elastic burst detection is innovative. We achieve efficient detection of subsequences with bursts in a time series by filtering lots of subsequences that are unlikely to have burst. This is an extension to the well-known lower bound technique in similarity search from time series [8].

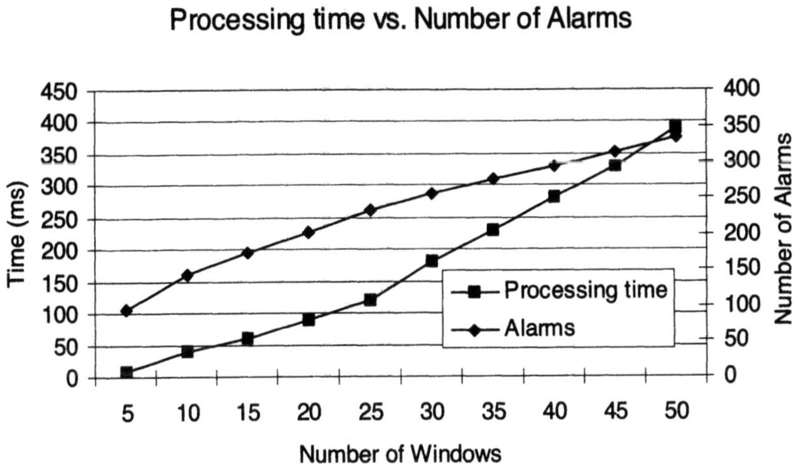

Figure 7.17. Processing time of elastic spread detection on Stock data for different output sizes.

Data mining on bursty behavior has attracted more attention recently. Wang et al. [95] study fast algorithms using self-similarity to model bursty time series. Such models can generate more realistic time series streams for testing data mining algorithm. Kleinberg [60] also discusses the problem of burst detection in data streams. The focus of his work is in modeling and extracting structure from text streams, decomposing time into a layered set of intervals corresponding to smaller and larger extents of bursty behavior. Our work is different in that we focus on the algorithmic issue of counting over different sliding windows, reporting all windows where the number of events exceeds a threshold based on the window length.

We have extended the data structure for burst detection to high-spread detection in time series. Spread measures the surprising level of time series. There is other work in finding surprising patterns in time series data. However the definition of surprise is application dependent and it is up to the domain experts to choose the appropriate one for their application. Jagadish et al. [49] use optimal histograms to mine deviants in time series. In their work deviants are defined as points with values that differ greatly from those of surrounding points. Shahabi et al. [87] also use a wavelet-based structure (TSA-Tree) to find both trends and surprises in large time series dataset, where surprises are defined as large difference between two consecutive averages of a time series. In very recent work, Keogh et al. [58] propose an algorithm based on suffix tree structures to find surprising patterns in time series database. They try to learn a model from the previously observed time series and declare surprising those patterns with small chance of occurrence. By comparison, an advantage of our definition of surprise based on spread is that it is simple, intuitive and scalable to massive and streaming data.

7.6 Conclusions and Future Work

This chapter introduces the concept of monitoring data streams based on an elastic window model and demonstrates the desirability of the new model. The beauty of the model is that the sliding window size is left for the system to discover in data stream monitoring. We also propose a novel data structure for efficient detection of elastic bursts and other aggregates. Experiments of our system OmniBurst on real data sets show that our algorithm is faster than a brute force algorithm by several orders of magnitude. We are currently collaborating with physicists to deploy our algorithm for online Gamma Ray burst detection. Future work includes:

- A robust way of setting the thresholds of burst for different window sizes;
- Algorithms to monitor non-monotonic aggregates such as medians;
- An efficient way to monitor the bursts of many different event types at the same time.

7.7 Questions

1. Suppose that we build a Shifted Binary Tree for a time series $\mathbf{x} = (x_1, x_2, \ldots, x_{1024})$, and the subsequence $(x_{100}, x_{101}, .., x_{120})$ experiences a burst, which window in the Shifted Wavelet Tree will report the burst?
2. For the Shifted Wavelet Tree, if the sizes of window increase 3 times as the levels increase, what should be the portion of overlapping at the same level to guarantee that the tree can still cover any window size?
3. In Lemma 7.4, what is the worst case processing time for insertion into the data stream for the batch algorithm.
4. Suppose you were interested in all window sizes up to 16 in a burst detection setting. Suppose that the thresholds for the window sizes were:

size	—	thresh
1	—	4
2	—	4
3	—	4
4	—	4
5	—	4
6	—	4
7	—	5
8	—	5
9	—	5
10	—	6
11	—	6
12	—	6
13	—	6
14	—	6
15	—	7
16	—	7

With which thresholds would you design the shifted wavelet tree?

5. Suppose that we build a two-dimensional Shifted Binary Tree for a grid of [1..64] × [1..64], what is the spatial window that will report the burst in the region [50..64] × [10..30]?

A Call to Exploration

Algorithmic improvements will be essential to time series analysis in the coming years. This might surprise those who marvel at the technology trends in which processor speed improvements of several orders of magnitude will occur in the coming decades. But there is no contradiction. Improvements in processors speed up existing algorithms, to be sure. But they also make detectors far more capable.

Stock prices and bids can be recorded and distributed from scores of markets in real time. Satellites and other spacecraft sport multi-frequency and multi-angle detectors of high precision. Single satellites may soon deliver trillions of time values per day. Less massive individually, but equally massive in the aggregate, magnetic resonance imagery machines report brain signals from tiny volume elements of the brain. These will soon be guiding brain surgery on a routine basis. Real-time feedback of a sophisticated kind may make new treatments possible.

What does this imply about algorithms?

- First, deriving useful knowledge will require fusing different time series data of the same type (e.g., time series from multiple voxels) or of different types (e.g., commodity prices and meteorological data).
- Second, filtering algorithms must be linear at worst and must do their job with high recall (few false negatives) and high precision (few false positives).
- Third, parameters (window sizes, thresholds, similarity measurements) will have to be inferred from the data and may change over time. To give one familiar example, stock volatilities have increased dramatically since the placid 1970s.

The primitives presented in this book both help to find interesting patterns in single time series (burst detection) and in multiple time series (correlation and distorted time search). If you gained a deep understanding from the material in the book, you will be able to invent new primitives for new applications.

In the current state of the art, primitives depend greatly on fixed parameters (thresholds, window sizes, lag values, and so on). They are nevertheless manifestly useful. But freeing these and similar primitives from the moorings of a priori parameter specifications – while satisfying the speed, recall, and precision constraints of data-rich applications – will open time series analysis to galactic possibilities.

Bon voyage.

A

Answers to the Questions

A.2 Chapter 2

2.
$$\begin{aligned}
DFT(x^*(-i)) &= DFT(x^*(N-i)) \\
&= \frac{1}{\sqrt{N}}\sum_{i=0}^{N-1} x^*(N-i)e^{-j2\pi Fi/N} \\
&= \frac{1}{\sqrt{N}}\sum_{k=1}^{N} x^*(k)e^{-j2\pi F(N-k)/N} \qquad (k = N-i) \\
&= \frac{1}{\sqrt{N}}\sum_{k=0}^{N-1} x^*(k)e^{-j2\pi F(N-k)/N} \qquad \text{(periodicity)} \\
&= \frac{1}{\sqrt{N}}\sum_{k=0}^{N-1} x^*(k)e^{-j2\pi F(-k)/N} \\
&= \frac{1}{\sqrt{N}}\sum_{k=0}^{N-1} x^*(k)e^{j2\pi Fk/N} \\
&= X^*(F)
\end{aligned}$$

4. Let $\mathbf{f}_2 = (1, 3, 5, 7, 4, 1, 5, 12, 17)$. Convolving $\mathbf{f}_2$ with the decomposition low-pass filter and downsampling the result gives

$$\mathbf{f}_1 = (-0.164, 3.207, 9.038, 2.336, 16.263, 8.210).$$

Convolving $\mathbf{f}_2$ with the decomposition high-pass filter and downsampling the result gives

$$\mathbf{d}_1 = (-0.612, 0, 0.647, -2.355, 10.884, -2.200).$$

Repeating the same process on $\mathbf{d}_1$, we have

$$\mathbf{f}_0 = (-0.452, 4.327, 8.902, 14.722),$$
$$\mathbf{d}_0 = (-1.686, 5.734, 7.946, -3.945).$$

The Inverse Discrete Wavelet Transform returns the original time series, possibly with zeroes padded at the end.

6. For $\mathbf{r}_j, j = 1, 2, \ldots, k$, we want to compute $\langle \mathbf{x}_1, \mathbf{r}_j^T \rangle$, $\langle \mathbf{x}_2, \mathbf{r}_j^T \rangle$, $\ldots$, $\langle \mathbf{x}_{N-n+1}, \mathbf{r}_j^T \rangle$. We know that the sequence of inner products can be computed from the convolution between $\mathbf{x}$ and $\tilde{\mathbf{r}}_j$ (obtained from $\mathbf{r}_j$ by padding zeroes and reversing the vector). The convolution can be computed from the product of their DFTs. The DFT and IDFT take time $O(N \log N)$, the product takes time $O(N)$. For each $\mathbf{r}_j, j = 1, 2, \ldots, k$, computing the sequence of inner products takes time $O(N \log N)$. The total time for the sketches computation is therefore $O(kN \log N)$.

A.3 Chapter 3

2. The KD-tree is a balanced tree. The quadtree is not a balanced tree.
4. There will be $(\frac{1}{0.05})^8 = 2.56 \times 10^{10}$ cells. We need to allocate 102.4GB memory for these cells. Therefore, the grid structure scales poorly to large numbers of dimensions. A dense representation of the grid structure consists of a hash structure. To perform a query, enter with a set of cell coordinates. If that isn't present in the hash structure, the cell is empty. If it is, the cell can be found by following a pointer from the hash structure.

A.4 Chapter 4

2. "Recall" defines the number of true positives returned by an approximate answer as a fraction of the size of the correct answer. "Precision" defines the number of true positives returned by an approximate answer as a fraction of the size of the approximate answer.
   In Question 1, the true positives are a, b, g, h, o, p, x, while the false positives are c, i, j and the false negatives are d, k. We have

$$\text{Recall} = \frac{\text{Number of True Positives}}{\text{Number of the Collect Answers}} = \frac{7}{9}$$

$$\text{Precision} = \frac{\text{Number of True Positives}}{\text{Size of the Approximate Answers}} = \frac{7}{10}.$$

4. For $F = 0, 1, \ldots, t-1$, we have

$$X(F) = \frac{1}{\sqrt{n}} \sum_{i=0}^{n-1} x(i) e^{-j2\pi Fi/n}$$

$$
\begin{aligned}
Y(F) &= \frac{1}{\sqrt{nw}} \sum_{k=0}^{nw-1} y(k) e^{-j2\pi Fk/nw} \\
&= \frac{1}{\sqrt{nw}} \sum_{i=0}^{n-1} \sum_{m=0}^{w-1} y(iw+m) e^{-j2\pi F(iw+m)/nw} \qquad (k = iw + m) \\
&= \frac{1}{\sqrt{nw}} \sum_{m=0}^{w-1} \sum_{i=0}^{n-1} x(i) e^{-j2\pi Fi/n} e^{-j2\pi Fm/nw} \\
&= \frac{1}{\sqrt{w}} \sum_{m=0}^{w-1} X(F) e^{-j2\pi Fm/nw} \\
&= X(F) \frac{1}{\sqrt{w}} \sum_{m=0}^{w-1} e^{-j2\pi Fm/nw}.
\end{aligned}
$$

A.5 Chapter 5

2. For $m = 1, \ldots, n$, we have

$$
X_m^{\text{old}} = \frac{1}{\sqrt{w}} \sum_{i=0}^{w-1} x_i e^{-j2\pi mi/w}
$$

$$
X_m^{\text{new}} = \frac{1}{\sqrt{w}} \sum_{i=0}^{w-1} x_{i+1} e^{-j2\pi mi/w}.
$$

Therefore,

$$
\begin{aligned}
e^{-j2\pi m/w} X_m^{\text{new}} &= \frac{1}{\sqrt{w}} \sum_{i=0}^{w-1} x_{i+1} e^{-j2\pi m(i+1)/w} \\
&= \frac{1}{\sqrt{w}} \left(\sum_{i=0}^{w-1} x_i e^{-j2\pi mi/w} + x_w e^{-j2\pi mw/w} - x_0 e^{-j2\pi m0/w} \right) \\
&= \frac{1}{\sqrt{w}} \left(\sum_{i=0}^{w-1} x_i e^{-j2\pi mi/w} + x_w - x_0 \right) \\
&= X_m^{\text{new}} + \frac{1}{\sqrt{w}} (x_w - x_0).
\end{aligned}
$$

4. We have

$$\sum_{i=1}^{b} f_m(i) f_p^*(i) = \frac{1}{\sqrt{b}} \sum_{i=1}^{b} e^{j2\pi mi/b} e^{-j2\pi pi/b}$$
$$= \frac{1}{\sqrt{b}} \sum_{i=1}^{b} e^{j2\pi(m-p)i/b}$$
$$= \begin{cases} 0 & m \neq p \\ \sqrt{b} \neq 0 & m = p. \end{cases}$$

6. (Hint) Using FFT to compute the convolution.

A.6 Chapter 6

2. Piecewise Aggregate Approximation gives a tighter envelope in most cases. Intuitively, the reason is that all the coefficients of the linear transformation for Piecewise Aggregate Approximation are positive, while some of the coefficients of the linear transformation for the Discrete Haar Wavelet Transform are not. Unfortunately, this is not a guarantee which is why numerical experiments are necessary.
4. To put an envelope around both time series in the database and the query time series is unnecessary. To put the envelope around the time series in the database does not give us flexibility to adjust the warping width in Dynamic Time Warping. To put an envelope around the query time series is better. It allows Dynamic Time Warping with different warping widths, up to and including Euclidean distance (set the warping width to 0).
6. No.

A.7 Chapter 7

2. 2/3.
4. The key point to note is that the threshold for a node in the shifted binary tree must be low enough to capture any interval not caught by lower nodes in the tree.

node	—	alarm thresh
2	—	4
4	—	4
8	—	4 (to cover the case where a window of size 5 has 4 events)
16	—	5
32	—	7

References

1. http://www.kx.com.
2. http://www.macho.mcmaster.ca/project/overview/status.html.
3. http://www.mathworks.com/.
4. http://www.netlib.org/lapack/.
5. http://www.nyse.com/taq/.
6. http://www.lanl.gov/milagro/, 2002.
7. D. Achlioptas. Database-friendly random projections. In *Proceedings of the Twentieth ACM SIGMOD-SIGACT-SIGART Symposium on Principles of Database Systems*, pages 274–281. ACM Press, 2001.
8. R. Agrawal, C. Faloutsos, and A. N. Swami. Efficient Similarity Search In Sequence Databases. In D. Lomet, editor, *Proceedings of the 4th International Conference of Foundations of Data Organization and Algorithms (FODO)*, pages 69–84, Chicago, Illinois, 1993. Springer Verlag.
9. AKoff Sound Labs. [http://www.akoff.com/music-composer.html] *Akoff Music Composer Version 2.0*, 2000.
10. A. Arasu, B. Babcock, S. Babu, M. Datar, K. Ito, I. Nishizawa, J. Rosenstein, and J. Widom. Stream: The Stanford stream data manager. In A. Y. Halevy, Z. G. Ives, and A. Doan, editors, *Proceedings of the 2003 ACM SIGMOD International Conference on Management of Data, San Diego, California, USA, June 9–12, 2003*, page 665. ACM, 2003.
11. B. Babcock, S. Babu, M. Datar, R. Motwani, and J. Widom. Models and issues in data stream systems. In *Proceedings of the Twenty-first ACM SIGACT-SIGMOD-SIGART Symposium on Principles of Database Systems, June 3–5, Madison, Wisconsin, USA*, pages 55–68. ACM, 2002.
12. B. Babcock, M. Datar, R. Motwani, and L. O'Callaghan. Maintaining variance and k-medians over data stream windows. In *PODS 2003, Proceedings of the Twenty-Second ACM SIGACT-SIGMOD-SIGART Symposium on Principles of Database Systems, June 9–12, 2003, San Diego, CA, USA*, pages 234–243. ACM, 2003.
13. S. Babu and J. Widom. Continuous queries over data streams. *SIGMOD Record*, 30(3):109–120, 2001.
14. D. Barbara, W. DuMouchel, C. Faloutsos, P. J. Haas, J. M. Hellerstein, Y. E. Ioannidis, H. V. Jagadish, T. Johnson, R. T. Ng, V. Poosala, K. A. Ross, and K. C. Sevcik. The New Jersey data reduction report. *Data Engineering Bulletin*, 20(4):3–45, 1997.

15. N. Beckmann, H.-P. Kriegel, R. Schneider, and B. Seeger. The r*-tree: An efficient and robust access method for points and rectangles. In *Proceedings of the 1990 ACM SIGMOD International Conference on Management of Data, Atlantic City, NJ, May 23–25, 1990*, pages 322–331, 1990.
16. J. L. Bentley, B. W. Weide, and A. C. Yao. Optimal expected-time algorithms for closest point problems. *ACM Transactions on Mathematical Software (TOMS)*, 6(4):563–580, 1980.
17. D. Berndt and J. Clifford. Using dynamic time warping to find patterns in time series. In *Advances in Knowledge Discovery and Data Mining*, pages 229–248. AAAI/MIT, 1994.
18. S. G. Blackburn and D. C. DeRoure. A tool for content based navigation of music. In *ACM Multimedia 98*, pages 361–368, 1998.
19. D. Carney, U. Cetintemel, M. Cherniack, C. Convey, S. Lee, G. Seidman, M. Stonebraker, N. Tatbul, and S. B. Zdonik. Monitoring streams – a new class of data management applications. In *VLDB 2002, Proceedings of 28th International Conference on Very Large Data Bases, August 20–23, 2002, Hong Kong, China*, 2002.
20. K. Chakrabarti, M. N. Garofalakis, R. Rastogi, and K. Shim. Approximate query processing using wavelets. In *VLDB 2000, Proceedings of 26th International Conference on Very Large Data Bases, September 10–14, 2000, Cairo, Egypt*, pages 111–122, 2000.
21. K.-P. Chan and A. W.-C. Fu. Efficient time series matching by wavelets. In *Proceedings of the 15th International Conference on Data Engineering, Sydney, Australia*, pages 126–133, 1999.
22. E. Cohen and M. Strauss. Maintaining time-decaying stream aggregates. In *PODS 2003, Proceedings of the Twenty-Second ACM SIGACT-SIGMOD-SIGART Symposium on Principles of Database Systems, June 9–12, 2003, San Diego, CA, USA*, pages 223–233. ACM, 2003.
23. J. W. Cooley and J. W. Tukey. An algorithm for the machine calculation of complex fourier series. *Mathematical Computations*, 19, April 1965.
24. G. Cormode, M. Datar, P. Indyk, and S. Muthukrishnan. Comparing data streams using hamming norms (how to zero in). In *VLDB 2002, Proceedings of 28th International Conference on Very Large Data Bases, August 20–23, 2002, Hong Kong, China*, 2002.
25. G. Cormode, P. Indyk, N. Koudas, and S. Muthukrishnan. Fast mining of massive tabular data via approximate distance computations. In *ICDE 2002, 18th International Conference on Data Engineering, February 26–March 1, 2002, San Jose, California*, 2002.
26. C. Cortes, K. Fisher, D. Pregibon, and A. Rogers. Hancock: A language for extracting signatures from data streams. In *ACM SIGKDD Intl. Conf. on Knowledge Discovery and Data Mining*, pages 9–17, 2000.
27. A. Das, J. Gehrke, and M. Riedewald. Approximate join processing over data streams. In A. Y. Halevy, Z. G. Ives, and A. Doan, editors, *Proceedings of the 2003 ACM SIGMOD International Conference on Management of Data, San Diego, California, USA, June 9–12, 2003*, pages 40–51. ACM, 2003.
28. T. Dasu, T. Johnson, S. Muthukrishnan, and V. Shkapenyuk. Mining database structure; or, How to build a data quality browser. In *Proceedings of the 2002 ACM SIGMOD International Conference on Management of Data*, pages 240–251. ACM Press, 2002.
29. M. Datar, A. Gionis, P. Indyk, and R. Motwani. Maintaining stream statistics over sliding windows. In *Proceedings of the Thirteenth Annual ACM-SIAM Symposium on Discrete Algorithms, January 6–8, 2002, San Francisco, CA, USA. ACM/SIAM, 2002*, pages 635–644, 2002.
30. I. Daubechies. *Ten Lectures on Wavelets*. SIAM, 1992.

31. A. Dobra, M. Garofalakis, J. Gehrke, and R. Rastogi. Processing complex aggregate queries over data streams. In *Proceedings of the 2002 ACM SIGMOD International Conference on Management of Data*, pages 61–72. ACM Press, 2002.
32. P. Domingos and G. Hulten. Mining high-speed data streams. In *Proceedings of the Sixth ACM SIGKDD International Conference on Knowledge Discovery and Data Mining*, pages 71–80. ACM Press, 2000.
33. R. Atkins et. al. (The Milagro Collaboration). Evidence for TeV emission from GRB 970417a. In *Ap.J. Lett. 533, L119*, 2000.
34. C. Faloutsos, M. Ranganathan, and Y. Manolopoulos. Fast subsequence matching in time-series databases. In *Proc. ACM SIGMOD International Conf. on Management of Data*, pages 419–429, 1994.
35. V. Ganti, J. Gehrke, and R. Ramakrishnan. Demon: Data evolution and monitoring. In *Proceedings of the 16th International Conference on Data Engineering, San Diego, California*, 2000.
36. J. Gehrke, F. Korn, and D. Srivastava. On computing correlated aggregates over continual data streams. In *Proc. ACM SIGMOD International Conf. on Management of Data*, 2001.
37. A. Ghias, J. Logan, D. Chamberlin, and B. C. Smith. Query by humming: Musical information retrieval in an audio database. In *ACM Multimedia 1995*, pages 231–236, 1995.
38. A. C. Gilbert, Y. Kotidis, S. Muthukrishnan, and M. Strauss. Surfing wavelets on streams: One-pass summaries for approximate aggregate queries. In *VLDB 2001*, pages 79–88. Morgan Kaufmann, 2001.
39. D. Q. Goldin and P. C. Kanellakis. On similarity queries for time-series data: Constraint specification and implementation. In *Proceedings of the 1st International Conference on Principles and Practice of Constraint Programming (CP'95)*, 1995.
40. N. Golyandina, V. Nekrutkin, and A. Zhigljavsky. *Analysis of Time Series Structure SSA and Related Techniques*. CHAPMAN and HALL/CRC, 2001.
41. M. Greenwald and S. Khanna. Space-efficient online computation of quantile summaries. In *Proc. ACM SIGMOD International Conf. on Management of Data*, pages 58–66, 2001.
42. S. Guha, N. Mishra, R. Motwani, and L. O'Callaghan. Clustering data streams. In *the Annual Symposium on Foundations of Computer Science, IEEE*, 2000.
43. A. Guttman. R-trees: A dynamic index structure for spatial searching. In B. Yormark, editor, *SIGMOD'84, Proceedings of Annual Meeting, Boston, Massachusetts, June 18–21, 1984*, pages 47–57. ACM Press, 1984.
44. J. M. Hellerstein, J. F. Naughton, and A. Pfeffer. Generalized search trees for database systems. In U. Dayal, P. M. D. Gray, and S. Nishio, editors, *Proc. 21st Int. Conf. Very Large Data Bases, VLDB*, pages 562–573. Morgan Kaufmann, 11–15 1995.
45. G. Hulten, L. Spencer, and P. Domingos. Mining time-changing data streams. In *Proceedings of the Seventh ACM SIGKDD International Conference on Knowledge Discovery and Data Mining*, pages 97–106. ACM Press, 2001.
46. P. Indyk. Stable distributions, pseudorandom generators, embeddings and data stream computation. In *40th Symposium on Foundations of Computer Science*, 2000.
47. P. Indyk, N. Koudas, and S. Muthukrishnan. Identifying representative trends in massive time series data sets using sketches. In *VLDB 2000, Proceedings of 26th International Conference on Very Large Data Bases, September 10–14, 2000, Cairo, Egypt*, pages 363–372. Morgan Kaufmann, 2000.

48. A. J. Smith for the Milagro Collaboration. A search for bursts of Tev gamma rays with Milagro. In *Proceedings of the 27th International Cosmic Ray Conference(ICRC 2001), August 7–15, 2001, Hamburg, Germany*, 2001.
49. H. V. Jagadish, N. Koudas, and S. Muthukrishnan. Mining deviants in a time series database. In M. P. Atkinson, M. E. Orlowska, P. Valduriez, S. B. Zdonik, and M. L. Brodie, editors, *VLDB'99, Proceedings of 25th International Conference on Very Large Data Bases, September 7–10, 1999, Edinburgh, Scotland, UK*, pages 102–113. Morgan Kaufmann, 1999.
50. J.-S. R. Jang and H.-R. Lee. Hierarchical filtering method for content-based music retrieval via acoustic input. In *Proceedings of the Ninth ACM International Conference on Multimedia*, pages 401–410. ACM Press, 2001.
51. T. Johnson and D. Shasha. Utilization of b-trees with inserts, deletes and modifies. In *Proceedings of the Eighth ACM SIGACT-SIGMOD-SIGART Symposium on Principles of Database Systems, March 29–31, 1989, Philadelphia, Pennsylvania*, pages 235–246. ACM Press, 1989.
52. W. B. Johnson and J. Lindenstrauss. Extensions of lipshitz mapping into Hilbert space. *Contemp. Math.*, 26:189–206, 1984.
53. E. Keogh. Exact indexing of dynamic time warping. In *VLDB 2002, Proceedings of 28th International Conference on Very Large Data Bases, August 20–23, 2002, Hong Kong, China*, pages 406–417, 2002.
54. E. Keogh, K. Chakrabarti, S. Mehrotra, and M. J. Pazzani. Locally adaptive dimensionality reduction for indexing large time series databases. In *Proc. ACM SIGMOD International Conf. on Management of Data*, 2001.
55. E. Keogh, K. Chakrabarti, M. Pazzani, and S. Mehrotra. Dimensionality reduction for fast similarity search in large time series. In *Databases. Knowledge and Information Systems 3(3)*, pages 263–286, 2000.
56. E. Keogh and T. Folias. [http://www.cs.ucr.edu/~eamonn/tsdma/index.html], The UCR time series data mining archive, Riverside, CA, University of California – Computer Science and Engineering Department, 2002.
57. E. Keogh and S. Kasetty. On the need for time series data mining benchmarks: A survey and empirical demonstration. In *The 8th ACM SIGKDD International Conference on Knowledge Discovery and Data Mining, July 23–26, 2002. Edmonton, Alberta, Canada*, pages 102–111, 2002.
58. E. Keogh, S. Lonardi, and B. Y. Chiu. Finding surprising patterns in a time series database in linear time and space. In *The 8th ACM SIGKDD International Conference on Knowledge Discovery and Data Mining, July 23–26, 2002. Edmonton, Alberta, Canada*, pages 550–556, 2002.
59. E. Keogh and P. Smyth. A probabilistic approach to fast pattern matching in time series databases. In *The Third conference on Knowledge Discovery in Databases and Data Mining*, 1997.
60. J. Kleinberg. Bursty and hierarchical structure in streams. In *The 8th ACM SIGKDD International Conference on Knowledge Discovery and Data Mining, July 23–26, 2002. Edmonton, Alberta, Canada*, pages 91–101, 2002.
61. F. Korn, H. V. Jagadish, and C. Faloutsos. Efficiently supporting ad hoc queries in large datasets of time sequences. In *SIGMOD 1997, Proceedings ACM SIGMOD International Conference on Management of Data, May 13–15, 1997, Tucson, Arizona, USA*, pages 289–300, 1997.
62. F. Korn, S. Muthukrishnan, and Y. Zhu. Checks and balances: Monitoring data quality problems in network traffic databases. In J. C. Freytag, P. C. Lockemann, S. Abiteboul,

M. J. Carey, P. G. Selinger, and A. Heuer, editors, *VLDB 2003, Proceedings of 29th International Conference on Very Large Data Bases, September 9–12, 2003, Berlin, Germany*, pages 536–547. Morgan Kaufmann, 2001.
63. F. Korn, S. Muthukrishnan, and Y. Zhu. Ipsofacto: A visual correlation tool for aggregate network traffic data. In A. Y. Halevy, Z. G. Ives, and A. Doan, editors, *Proceedings of the 2003 ACM SIGMOD International Conference on Management of Data, San Diego, California, USA, June 9–12, 2003*, page 677. ACM, 2003.
64. F. Korn, N. Sidiropoulos, C. Faloutsos, E. Siegel, and Z. Protopapas. Fast nearest neighbor search in medical image databases. In *VLDB'96, Proceedings of 22th International Conference on Very Large Data Bases, September 3–6, 1996, Mumbai (Bombay), India*, pages 215–226, 1996.
65. N. Kosugi, Y. Nishihara, T. Sakata, M. Yamamuro, and K. Kushima. A practical query-by-humming system for a large music database. In *ACM Multimedia 2000*, pages 333–342, 2000.
66. C.-S. Li, P. S. Yu, and V. Castelli. Hierarchyscan: A hierarchical similarity search algorithm for databases of long sequences. In *ICDE*, pages 546–553, 1996.
67. S. Mallat. *A Wavelet Tour of Signal Processing*. Academic Press, 1998.
68. G. S. Manku, S. Rajagopalan, , and B. G. Lindsay. Random sampling techniques for space efficient online computation of order statistics of large datasets. In *Proc. ACM SIGMOD International Conf. on Management of Data*, pages 251–262, 1999.
69. Y. Matias, J. S. Vitter, and M. Wang. Wavelet-based histograms for selectivity estimation. In L. M. Haas and A. Tiwary, editors, *SIGMOD 1998, Proceedings ACM SIGMOD International Conference on Management of Data, June 2–4, 1998, Seattle, Washington, USA*, pages 448–459, 1998.
70. D. Mazzoni and R. B. Dannenberg. Melody matching directly from audio. In *2nd Annual International Symposium on Music Information Retrieval, Bloomington, Indiana, USA*, 2001.
71. R. J. McNab, L. A. Smith, D. Bainbridge, and I. H. Witten. The New Zealand digital library melody index. In *D-Lib Magazine*, 1997.
72. M. Misti, Y. Misti, G. Oppenheim, and J. Poggi. *Wavelet Toolbox User's Guide*. Math Works Inc., Massachusetts, 1996.
73. L. Molesky and M. Caruso. Managing financial time series data: Object-relational and object database systems. In *VLDB'98, Proceedings of 24rd International Conference on Very Large Data Bases*, 1998.
74. Y.-S. Moon, K.-Y. Whang, and W.-S. Han. General match: A subsequence matching method in time-series databases based on generalized windows. In *Proceedings of the 2002 ACM SIGMOD International Conference on Management of data*, pages 382–393. ACM Press, 2002.
75. S. Muthukrishnan. Data streams: Algorithms and applications. In *Proceedings of the Fourteenth Annual ACM-SIAM Symposium on Discrete Algorithms*, pages 413–413. Society for Industrial and Applied Mathematics, 2003.
76. J. Nolan. An introduction to stable distributions.
77. S. Park, W. W. Chu, J. Yoon, and C. Hsu. Fast retrieval of similar sub-sequences under time warping. In *ICDE*, pages 23–32, 2000.
78. I. Popivanov and R. J. Miller. Similarity search over time series data using wavelets. In *ICDE*, 2002.
79. W. Press, B. Flannery, S. Teukolsky, and W. Vetterling. *Numerical Recipes: The Art of Scientific Computing*. Cambridge University Press, 1986.
80. J. Profita and T.G. Bidder. Perfect pitch. In *American Journal of Medical Genetics*, pages 763–771, 1988.

81. D. Rafiei and A. Mendelzon. Similarity-based queries for time series data. In *Proc. ACM SIGMOD International Conf. on Management of Data*, pages 13–25, 1997.
82. D. Rafiei and A. Mendelzon. Efficient retrieval of similar time sequences using dft. In *Proc. FODO Conference, Kobe, Japan*, 1998.
83. H. Samet. The quadtree and related hierarchical data structures. *ACM Computing Surveys*, 16(2):187–260, 1984.
84. M. Schroeder. *Fractals, Chaos, Power Laws: Minutes From an Infinite Paradise*. W. H. Freeman and Company, New York, 1991.
85. T. Seidl and H.-P. Kriegel. Optimal multi-step k-nearest neighbor search. In L. M. Haas and A. Tiwary, editors, *SIGMOD 1998, Proceedings ACM SIGMOD International Conference on Management of Data, June 2–4, 1998, Seattle, Washington, USA*, pages 154–165, 1998.
86. T. K. Sellis, N. Roussopoulos, and C. Faloutsos. The r+-tree: A dynamic index for multi-dimensional objects. In P. M. Stocker, W. Kent, and P. Hammersley, editors, *VLDB'87, Proceedings of 13th International Conference on Very Large Data Bases, September 1–4, 1987, Brighton, England*, pages 507–518. Morgan Kaufmann, 1987.
87. C. Shahabi, X. Tian, and W. Zhao. Tsa-tree: A wavelet-based approach to improve the efficiency of multi-level surprise and trend queries on time-series data. In *12th International Conference on Scientific and Statistical Database Management (SSDBM'00), July 26–28, 2000, Berlin, Germany*, pages 55–68, 2000.
88. D. E. Shasha and P. Bonnet. *Database Tuning: Principles, Experiments, and Troubleshooting Techniques*. Morgan Kaufmann, 2002.
89. N. Thaper, S. Guha, P. Indyk, and N. Koudas. Dynamic multidimensional histograms. In *Proceedings of the 2002 ACM SIGMOD International Conference on Management of Data*, pages 428–439. ACM Press, 2002.
90. T. Tolonen and M. Karjalainen. A computationally efficient multi-pitch analysis model. *IEEE Transactions on Speech and Audio Processing*, 2000.
91. A. Uitdenbgerd and J. Zobel. Manipulation of music for melody matching. In *ACM Multimedia 98*, pages 235–240, 1998.
92. A. Uitdenbgerd and J. Zobel. Melodic matching techniques for large music databases. In *ACM Multimedia 99*, pages 57–66, 1999.
93. J. S. Vitter and M. Wang. Approximate computation of multidimensional aggregates of sparse data using wavelets. In *SIGMOD 1999, Proceedings ACM SIGMOD International Conference on Management of Data, June 1–3, 1999, Philadephia, Pennsylvania, USA*, pages 193–204, 1999.
94. D. F. Walnut. *An Introduction to Wavelet Analysis*. Birkhäuser, 2002.
95. M. Wang, T. M. Madhyastha, N. H. Chan, S. Papadimitriou, and C. Faloutsos. Data mining meets performance evaluation: Fast algorithms for modeling bursty traffic. In *ICDE 2002, 18th International Conference on Data Engineering, February 26–March 1, 2002, San Jose, California*, 2002.
96. H. J. Weaver. *Theory of Discrete and Continuous Fourier Analysis*. John Wiley & Sons, 1989.
97. Y.-L. Wu, D. Agrawal, and A. E. Abbadi. A comparison of DFT and DWT based similarity search in time-series databases. In *Proceedings of the 9th International Conference on Information and Knowledge Management*, 2000.
98. C. Yang. Efficient acoustic index for music retrieval with various degrees of similarity. In *ACM Multimedia 2002, December 1–6, 2002, French Riviera*, 2002.
99. B.-K. Yi and C. Faloutsos. Fast time sequence indexing for arbitrary LP norms. In *VLDB 2000, Proceedings of 26th International Conference on Very Large Data Bases, September 10–14, 2000, Cairo, Egypt*, pages 385–394. Morgan Kaufmann, 2000.

100. B.-K. Yi, H. V. Jagadish, and C. Faloutsos. Efficient retrieval of similar time sequences under time warping. In *ICDE*, pages 201–208, 1998.
101. B.-K. Yi, N. Sidiropoulos, T. Johnson, H. V. Jagadish, C. Faloutsos, and A. Biliris. On-line data mining for co-evolving time sequences. In *Proceedings of the 16th International Conference on Data Engineering, San Diego, California*, pages 13–22, 2000.
102. Y. Zhu, M. S. Kankanhalli, and C. Xu. Pitch tracking and melody slope matching for song retrieval. In *Advances in Multimedia Information Processing–PCM 2001, Second IEEE Pacific Rim Conference on Multimedia, Beijing, China, October 24–26*, 2001.
103. Y. Zhu and D. Shasha. Statstream: Statistical monitoring of thousands of data streams in real time. In *VLDB 2002, Proceedings of 28th International Conference on Very Large Data Bases, August 20–23, 2002, Hong Kong, China*, pages 358–369, 2002.
104. Y. Zhu and D. Shasha. Efficient elastic burst detection in data streams. In *KDD 2003, Proceedings of the Ninth ACM SIGKDD International Conference on Knowledge Discovery and Data Mining, August 24–27, 2003, Washington, DC, USA*. ACM, 2003.
105. Y. Zhu and D. Shasha. Fast approaches to simple problems in financial time series streams. In S. Muthukrishnan and D. Srivastava, editors, *MPDS 2003, Proceedings of the ACM SIGMOD/PODS and FCRC 2003 Workshop on Management and Processing of Data Streams, San Diego, California, USA, June 8, 2003*, pages 181–192. ACM, 2003.
106. Y. Zhu and D. Shasha. Warping indexes with envelope transforms for query by humming. In A. Y. Halevy, Z. G. Ives, and A. Doan, editors, *Proceedings of the 2003 ACM SIGMOD International Conference on Management of Data, San Diego, California, USA, June 9–12, 2003*, pages 181–192. ACM, 2003.
107. Y. Zhu, D. Shasha, and X. Zhao. Query by humming – in action with its technology revealed. In A. Y. Halevy, Z. G. Ives, and A. Doan, editors, *Proceedings of the 2003 ACM SIGMOD International Conference on Management of Data, San Diego, California, USA, June 9–12, 2003*, page 675. ACM, 2003.

Index

MIX
Papier aus verantwortungsvollen Quellen
Paper from responsible sources
FSC® C105338

If you have any concerns about our products,
you can contact us on
ProductSafety@springernature.com

In case Publisher is established outside the EU,
the EU authorized representative is:
Springer Nature Customer Service Center GmbH
Europaplatz 3, 69115 Heidelberg, Germany

Printed by Libri Plureos GmbH
in Hamburg, Germany